# Notes From the Editor

Sun Zi's *The Art of War*, written about 2,500 years ago, is rich in contents and profound in meaning and is honored as the Bible of military sciences.

In addition to its relevance to military matters, the philosophy and insights found in the work are widely used in enterprise management and commercial and sports competition. In China, they are also used in traditional Chinese medicine (TCM). For example, traditional Chinese doctors, such as Bian Que, over 2,000 years ago, Sun Simiao, 581-682, and Xu Dachun in the Qing Dynasty (1644-1911) proposed "conquering diseases like conquering enemies" and "using medicines like using soldiers".

Inheriting these theories, Wu Rusong, a contemporary expert on Sun Zi, Wang Hongtu, a professor of TCM, and Huang Ying, a doctor of TCM, wrote this book, the first in Chinese history explaining the relationship between TCM and Chinese military strategies.

The 43 chapters illustrate the relationships between military and medical sciences. With typical cases and modern ideas, the book is of use to both doctors and lay people.

Doctors, whether they practice Western medicine or TCM, will surely benefit from the application of military sciences to medicine.

*Wu Xianlin*

# Authors

Wu Rusong: Expert on Sun Zi's *The Art of War*; Doctoral Tutor

Wang Hongtu: Expert in Traditional Chinese Medicine; Doctoral Tutor

Huang Ying: Military Surgeon

# Sun Zi's Art of War and Health Care

**Translators**
She Duanzhi   Li Bin

**English Editors**
Richard R. Pearce   Robert Biddle
Anne-Marie Robinson   Geoff Murray

**Executive Editor**
Wang Yanjuan

NEW WORLD PRESS
BEIJING, CHINA

**First Edition: 1997**

Copyright by **New World Press**, Beijing, China.
All rights reserved. No part of this book may be reproduced in any form or by any means without permission in writing from the publisher.

Chief Editor: Wu Xianlin
Executive Editor: Chen Shiyue

**ISBN 7-80005-376-8**

*Published by*
NEW WORLD PRESS
24 Baiwanzhuang Road, Beijing 100037, China

*Distributed by*
CHINA INTERNATIONAL BOOK TRADING CORPORATION
35 Chegongzhuang Xilu, Beijing 100044, China
P.O. Box 399, Beijing, China

*Printed in the People's Republic of China*

# CONTENTS

**Remarks on Terminology** ............................................................ 1

**Foreword** ............................................................ 14

| | | |
|---|---|---|
| Chapter 1 | The Value of Humanity | 18 |
| Chapter 2 | Science vs. Superstition | 24 |
| Chapter 3 | Military Preparedness and Preventative Medicine | 30 |
| Chapter 4 | Prepare and Prevent | 37 |
| Chapter 5 | Military Morality and Self-Cultivation | 44 |
| Chapter 6 | Terrain and Topography | 48 |
| Chapter 7 | More Is Not Always Better | 52 |
| Chapter 8 | The Benefits of Exercise | 57 |
| Chapter 9 | Balance Work and Rest | 66 |
| Chapter 10 | Anger Equals Defeat | 80 |
| Chapter 11 | Mental Calm Promotes Health | 84 |
| Chapter 12 | Regularity Is Significant | 90 |
| Chapter 13 | Excessive Maintenance Is Bad | 98 |
| Chapter 14 | Sex Drive Good and Bad | 101 |
| Chapter 15 | Qigong and Morale | 108 |
| Chapter 16 | Choices Are Important | 124 |
| Chapter 17 | Solidarity | 128 |
| Chapter 18 | Observe the Situation | 131 |
| Chapter 19 | Analysis Important | 141 |
| Chapter 20 | Food vs. Medicine | 146 |
| Chapter 21 | The Significance of Tactics | 151 |
| Chapter 22 | Formulas and Formations | 155 |
| Chapter 23 | Flexibility | 159 |
| Chapter 24 | Discretion | 165 |
| Chapter 25 | The Heart of the Matter | 172 |
| Chapter 26 | Taming Emotions | 178 |
| Chapter 27 | Diversion | 182 |

| Chapter 28 | Dispelling Suspicions | 187 |
| Chapter 29 | Boosting Morale | 194 |
| Chapter 30 | Guidance | 197 |
| Chapter 31 | Removing the Root Causes | 208 |
| Chapter 32 | Bolstering Resistance | 213 |
| Chapter 33 | Preventing Disease | 217 |
| Chapter 34 | Peace of Mind | 220 |
| Chapter 35 | The Roots of Disease | 223 |
| Chapter 36 | Holistic Treatment | 226 |
| Chapter 37 | Understanding the Six Channels | 229 |
| Chapter 38 | People and the Army | 231 |
| Chapter 39 | Different Approaches to One Disease | 233 |
| Chapter 40 | One Approach to Different Diseases | 237 |
| Chapter 41 | Good Timing | 241 |
| Chapter 42 | Exploiting the Circumstances | 243 |
| Chapter 43 | Costs and Benefits | 246 |// 
**Text in Chinese** ......... 249

# Remarks on Terminology

Traditional Chinese medicine (TCM) took shape at roughly the same time as Sun Zi wrote *The Art of War*. Both were influenced by the philosophies of their time. This explains why they use many similar or identical terms in their discourse. These include *yin* and *yang*, *xu* (rendered as deficiency here) and *shi* (rendered as excesses here), *xie* (pathogenic factors) and *zheng* (body resistance). The 43 chapters in this book testify to the close relationship between medicine and warfare.

However, since the Chinese language has changed dramatically from what it was 2,000 years ago, today's readers may have problems understanding the classics. Therefore, brief explanatory notes on the terminology used in the book are perhaps merited.

**Yin and Yang**

A popular philosophy during the Spring and Autumn and Warring States Periods (770-221 BC) was the *yin-yang* school which interpreted the universe in terms of two opposing concepts of *yin* and *yang*. *The Yellow Emperor's Canon of Internal Medicine* defines *yin* and *yang* as universal constants, the key to understanding the world, the source of all changes and the inherent logic for the evolution of all things. In a word, *yin* and *yang* account for the evolution of the universe. It is, therefore, a very important concept in diagnosis and therapy. In warfare, likewise, the principle of *yin* and *yang* is instrumental in all maneuvers from camping to deployment.

This school encompasses the following components:

First, *yin* and *yang* are mutually opposed, restrictive and balanced. As mentioned above, *yin* and *yang* are omnipresent in the universe. Generally, anything bright, superior, active, full, superficial, hot, light and clear is considered *yang*; by contrast, anything dark, subordinate, quiet, weak, inherent, cold, heavy and turbid is considered *yin*. For example, heaven is high above, light and clear and is therefore considered *yang*, whereas earth is subordinate, heavy and turbid and is therefore considered *yin*. The sun is bright and hot and is called *Taiyang* (extreme *yang*), whereas the moon is cold and often incomplete and is therefore called *Taiyin* (extreme *yin*). Other examples: fire, *yang*, and water, *yin*; man, *yang*, and woman, *yin*.

Even within the same human body, the heart and the lungs are *yang* because they are located in the upper part of the body, whereas the liver, spleen and kidneys are *yin* because of their subordinate position. The *qi* of the body is invisible and active and is therefore *yang*, while the blood is visible and circulates within the pulse and is *yin*. There are 12 major channels on the human body, six of which run on the exterior of the body and are considered *yang*, whereas the other six run in the interior of the body and are considered *yin*. For every internal organ, the physical part is *yang*, and the functional part *yin*. This is why in TCM there are concepts such as heart-*yin*, heart-*yang*, kidney-*yin*, kidney-*yang*, spleen-*yin* and spleen-*yang*.

Because of their mutually opposing nature, *yin* and *yang*, of necessity, check and balance each other. Too much cold will inevitably results in insufficient heat and vice versa. Within a year, spring and summer are considered *yang* because the weather gets warmer daily; autumn and winter are considered *yin* because the heat is on the wane. If one increases, the other decreases.

A patient running a high fever, for example, often feels thirsty, and his skin is often wrinkled. The fever is considered "*yang* on the rise", while thirst and wrinkles are "*yin* on the decline". In therapy, attention should be paid to both clearing away the *yang*-heat and increasing body fluids. Sometimes the patient may appear to have "*yang* on the rise", but in fact it is caused by "*yin* on the decline". The deficient *yin* cannot balance the growth of *yang*. Therefore, therapy should focus on strengthening *yin* to subdue *yang* (heat).

Second, *yin* and *yang* are not just opposites; they are also unified and interchangeable. This has two implications. One is the basis for the existence of the other. Without one, the other does not exist. Cold and heat, for example, define each other; so do exterior and interior, failure and victory. The physical body (*yin*) is useless without *qi* (*yang*); likewise, without the physical body, *qi* has no abode of existence. Health in this sense means a flourishing state of both the physical body and *qi*, a coordinated balance between *yin* and *yang*. Fundamentally, therapy is the effort to balance the *yin* and *yang* elements of the body.

The other implication of the connection of *yin* and *yang* is that when one reaches the end, the other begins. This is true of natural phenomena as well as medicine. For example, an excess of *yang* (known as the *yang* syndrome), clinically observed as a high fever, a reddened face and a strong pulse, may turn into a *yin* syndrome characterized by low body temperature, a pale face and weak pulse, if therapy is not effective.

Third, *yin* and *yang* are relative. Generally, day is *yang*, and night is *yin*. However, during any given day, the morning is *yang*, and the afternoon is *yin*. During the night, the first half is *yin*, and the second *yang*. Heaven is *yang*, but it too contains *yin* elements that are necessary for the formation of rain and dew. Earth is *yin*, but it, too, carries some *yang* characteristics. The melding of *yin* and *yang* gives rise to all things.

**The Five Elements**

The Five Elements refer to wood, fire, earth, metal and water. These are philosophical concepts used by ancient Chinese to explain the relationships between things. The Five-Element Theory, as it is known, was also popular during the Spring and Autumn and Warring States Periods. In TCM, the theory is used chiefly to explain the properties of viscera of the human body, their mutual relations, physiological phenomena and pathological changes. It also serves as a guide for diagnosing and treating diseases. The properties of the five elements are:

**Wood:** Supple, reaching upward, flourishing and lively. The liver, for example, is a "wood" organ. It is the source of vigor for the human body. The tendons are also regarded as "wood" since they are supple and yet strong.

**Fire:** Hot, active and burning, like the scorching

sun in summer. An equivalent viscera is the heart, which pumps blood all the time. Another one is the tongue.

**Earth:** Plain, quiet, accommodating, nourishing, like the lush greenery at the end of summer. An equivalent viscera is the spleen, which absorbs nutrients from food to produce *qi*, blood and fluids that are essential for the operations of the body. A strong spleen is especially important for the muscles.

**Metal:** Cold, versatile, like an autumn breeze that brings down foliage. An equivalent viscera is the lungs, responsible for the respiratory system. An important function of the lungs is to keep the *qi* of the body centered downward; otherwise it will cause coughing.

**Water:** Calm and cold, tending to move downward. An equivalent viscera is the kidneys, home to the essence of reproduction. This essence must be stored in full so as to prevent osteoporosis.

Like *yin* and *yang*, the Five Elements both create and check one another. Wood creates fire; fire creates earth; earth creates metal; metal creates water; and water creates wood. The cycle is endless. Wood checks earth; earth checks water; water checks fire; fire checks metal; and metal checks wood. The process is never-ending.

The Five Elements are interconnected and interdependent. No single element is independent of the other four.

The *yin* and *yang* theory is a dialectic approach concerned with the unity of opposites, while the Five-Element theory is based on the interrelationships of things. Over the long course of history, however, these two schools intermingled with each other into a unified theory that was applied in such diversified areas as astronomy, geology, calendar making, fine arts, agriculture, water conservation, war and medicine. Even today, the theory is an integral part of TCM, as can be seen in the use of such common names as heart-*yin*, heart-*yang*, *yin*-collateral and *yang*-collateral. Even practitioners of Western medicine habitually refer to a positive test result as "*yang*" and a negative result as "*yin*".

**Xie and Zheng**

*Zheng* is a Chinese character meaning upright, just and correct. Anything beneficial to the society and human health is called "*zheng*". A war against invasion and oppression is also regarded as "*zheng*". In TCM, *zheng qi* can refer to both natural and human phenomena. In nature, wind, cold, summer heat, dampness, dryness and fire are called the "six *qi*" that, within a normal range, are essential for the survival of life. For this reason, they are regarded as "*zheng*" *qi*. Within the human body, *qi*, blood, thin body fluids, fluids, essence and spirit constitute the body's resistance mechanism and are regarded as the *"zheng"* qi of the body.

*Xie*, or evil, is usually used to describe anything harmful to the society and human health. In TCM, it refers to pathogenic factors, which can be either natural or human. The above-mentioned "six *qi*", if moving out of the normal range and becoming pathogenic, are considered "*xie*" or exopathogen. Overdrinking, indigestion or swallowing poison are a few examples of *xie qi*, or pathogenic factors.

The human body itself is capable of producing pathogenic factors. Overstrain and nervousness, for example, can cause a tightening of the chest, insomnia, toothache and constipation. In TCM, this is known as "fire", which is also a kind of *xie qi*, or pathogenic factor. The fire actually is a result of hyperactivity of some part of the body, resulting in an excess of *zheng qi*. In addition, because of a dysfunction of certain organs, a person can have a number of symptoms, such as phlegm retention, blood stasis and dampness, which are also pathogenic.

It follows, therefore, that *xie* and *zheng* are relative. They are actually the same thing. Within a certain range, something can be *zheng*, but once beyond that range, it becomes *xie*. Blood, for example, is one of the essentials of life and is therefore normally regarded as *zheng*. However, when it is stagnant, it becomes *xie* (pathogenic factor).

This relativity also manifests itself in individual differences. Mr X and Mr Y, for example, live in the same climate. Mr X is quite healthy, but Mr Y falls ill. The weather, in this case, is *zheng* for Mr X, but *xie* for Mr Y. The reason Mr X is immune to the weather is because the *zheng qi* (body resistance) within his body is strong, whereas that of Mr Y is weak. The key to health care, therefore, is to nourish the *zheng qi* of the body.

Sometimes, though, a person with sufficient *zheng qi* may also fall victim to fierce *xie qi*. Therefore, *The Yellow Emperor's Canon of Internal Medicine* advises that wherever possible, one should avoid pathogenic factors. One is much better off staying away from alcohol than seeking a detoxicating drug. Likewise, prevention is the best medicine for such deadly diseases as AIDS.

**Xu and Shi**

Like *zheng* and *xie*, *xu* and *shi* are also relative. *Shi* means excess, abundance and sthenia; *xu* means insufficiency, weakness, deficiency and asthenia. In warfare, a common strategy is to avoid the strong and attack the weak. Another is to use camouflage to deceive the enemy.

In TCM, there are two sets of standards for diagnosis. One is strong or weak, and the other is *xu* or *shi*. A *shi* syndrome (or sthenic syndrome) is characterized by an excess of *xie qi* or pathogenic factors. This can be exopathogens or internal blood stasis, phlegm or fire. The general principle of therapy for such syndromes is purgation. An exopathogenic disease affects primarily the skin and the exterior body parts, and in therapy, sweat-inducing medicines should be used. Toxics and indigestion affect the stomach and should be treated with vomit-inducing drugs. Constipation affects the intestines, and purgation should be used. Blood stasis should be treated with drugs to promote the flow of blood. Fire and heat dwell inside the body and should be relieved with heat-clearing and fire-dispelling medicines. Water causes dropsy and difficulty in urination and should be treated with urine-inducing medicines. All these involve purgation to rid the body of the excessive pathogenic factors.

The same is true of the organs. A heart fire, for example, results in aphthae and dark urine and should be treated with fire-purging medicines. A liver fire can cause irritation, dizziness, tinnitus and constipation and should be quenched with medicines bitter in taste and cold in nature. A kidney fire causes insomnia, seminal emission and hyperactive sexual desire and should be treated with "zhibo dihuang wan", an essence of anemarrhena, phellodendron and rehmannia to purge the excessive fire.

A *xu* syndrome is characterized by deficiencies of *zheng qi* or body resistance. A general principle for treating such diseases is to invigorate the body resistance. Lassitude, a pale complexion and a weak pulse are symptoms of a *qi*-deficiency syndrome and can be treated with "shenling baizhu wan", an essence of ginseng, poria and bighead atractylodes.

Likewise, a blood deficiency syndrome, whose symptoms include a sallow complexion, a tightening of the chest, insomnia, scanty menstruation and a weak pulse, should be treated with blood-promoting medicines, such as "siwu tang", a decoction of four ingredients, and donkey-hide gelatin. A *yin*-deficiency syndrome has symptoms such as a feverish sensation over the palms and soles, a reddened face, night sweats and a shallow pulse. This can be treated with "liuwei dihuang wan", a bolus of six drugs, including rehmannia. A *yang*-deficiency syndrome, on the other hand, is characterized by cold limbs, thin defecation, clear urine and a sunken pulse. It can be treated with kidney-invigorating drugs. The internal organs can also suffer from deficiencies. The doctor must discern clinically which organ is affected and whether the deficiency is of *qi* or blood, *yin* or *yang*. A heart deficiency syndrome, for example, can be of heart-*qi*, heart-blood, heart-*yin* or heart-*yang*. Only when the location and origin of deficiency is identified can medicines be administered.

Despite the great numbers of *xu* and *shi* syndromes, an experienced doctor has no difficulty distinguishing one from another. Some patients, however, suffer from both a deficiency of *zheng qi* (body resistance) and an excess of *xie qi* (pathogenic factors). Under such circumstances, it is difficult to tell which element prevails. Although the general principle for treating such diseases is "combining purgation with invigoration", clinically it is difficult to decide on the proportion of purgatives to invigoratives and the sequence of drug administration. Take cancer for example. There is no doubt that the malignant tumor represents an excess of sthenic factors that should be removed through purgation. However, the patient's *zheng qi* (body resistance) is already weak and should be strengthened. Purgatives, unfortunately, have the side effect of weakening body resistance. Tonics, on the other hand, may aid pathogenic factors. The key to therapy, therefore, is to find the right balance between purgation and invigoration.

Some diseases are even more puzzling than a combination of *xu* and *shi* syndromes. These diseases can be quite deceptive: a sthenic disease may appear asthenic, and an asthenic disease may appear sthenic. It is important that the doctor not be deceived by appearances. A woman with blood stasis (sthenia) may instead have asthenic symptoms such as stoppage of menstruation, skinniness and dry skin. It would be a terrible mistake to treat her as asthenic. The correct way is to prescribe blood stasis-breaking medicines such as dahuang, or dried root and rhizome of rhubarb.

To the discerning doctor, there are minor symptoms behind the deceptive appearances that indicate the real disease. The woman with blood stasis mentioned above, for example, must have purplish spots on her tongue, and her pulse must be strong and sthenic.

This does not mean, of course, that misdiagnosis will never occur. In fact, even with the aid of modern medical equipment, misdiagnosis still happens from time to time.

# Foreword

In the mid-18th century, a doctor by the name of Xu Dachun rose to fame from Wujiang in Jiangsu Province. He once practiced medicine in the imperial hospital. He wrote many medical books illustrating his medical theories. Included in his works was the astute observation, "treatments for every disease are contained in the 13 chapters of Sun Zi's *The Art of War*".

This valuable and enlightening conclusion was drawn by Xu from his research in medical and military theories.

Even earlier, *The Yellow Emperor's Canon of Internal Medicine* had commented on the relation between medicine and the military.

And Sun Simiao further pointed out that a doctor who faced a patient was like a soldier facing his enemy. He had to evaluate the situation and make a judgement based on the fullest information about the enemy and his own forces. Then he should act accordingly.

Sun Zi stated that in war, tactics must be flexible

so that victory can be realized. And when treating a patient, the doctor must use medications according to concrete conditions.

The theories and strategies of military science can be applied to medical science. This has become a common belief of both ancient and modern Chinese doctors. So what are the connections between the two?

First, TCM considers medical science and skills as benevolent. Zhang Zhongjing, a doctor who lived about 2,000 years ago, said that doctors should provide medical treatment and services to ordinary citizens just as he would to emperors and his family members. The spirit of this idea is the same in military science. Military scientists say, "put down riots and eliminate injustice". That's why Sun Zi required military commanders to "advance without thinking of gaining personal fame, withdraw without fear of punishment, protect the citizenry and serve his sovereign".

Second, medical and military sciences both have complete and compact theoretical systems. For example, one ancient physician once said that excessive weather, whether fine, cloudy, windy and raining, is a factor in the cause of illnesses.

Applying treatment according to concrete conditions is the basic principle of TCM. Chinese doctors consider the sky, the earth and human beings as parts of a single system, and in the human body are many branch systems. Having made a diagnosis by means of looking, listening, questioning and feeling the pulse, a doctor should use medication to eliminate illness and improve weak organs.

Military science also supports systematized analysis. *The Art of War* calls for a general knowledge of the overall situation of the war and a grasp of the changing relations between time, terrain and human beings.

This theory of system is actually the cream of all Chinese academic theories. It is essentially different from Western philosophies. Traditional Chinese academic thoughts emphasize wholeness, spontaneity, harmony and coordination.

And third, there are many common points in medical and military sciences.

1. Combatting disease is like combatting the enemy. TCM believes that a good doctor knows well how to prevent an illness from ever occurring. And military scientists support keeping order before riots and rebellion can occur and being ready for sudden incidents so that "you can never fail".

2. Choosing a doctor is like selecting a military commander.

The ancient Chinese said, "Entrust your army to him if you know his wisdom. And entrust your life to him if you know his medications and skills."

A good commander should be selected to conduct

war, and a good doctor be chosen to treat diseases. The reasons are the same.

3. Applying medicines is like using military forces. Military forces are cruel, dangerous and violent. So are medicines. Therefore, a doctor must be careful in prescribing medications for his patients.

Ancient Chinese doctors also believed, "treating diseases is like fighting the enemy. A good commander wins a battle with proper attack and defense. And a good doctor cures a patient with proper method and medicines."

Military scientists also said, "a good commander uses his forces like a good doctor using medicines. Medicines are changed when the disease changes."

Thus, it can be seen that military theories are also enlightening to medical sciences. The important thing is to have a full understanding. That is to learn and practice so that knowledge can be enriched, and skills be made more exquisite.

## Chapter 1

## The Value of Humanity

Sun Bin, a Chinese military strategist living in the Warring States Period (475-221 BC), said in *Sun Bin's The Art of War*, "Human beings are the most valuable in the world." This shows how much ancient military scientists valued human life.

*The Yellow Emperor's Canon of Internal Medicine*, an ancient book on traditional Chinese medical sciences, says, "Nothing on the earth is more valuable than human beings." This shows the love medical scientists had for human life.

One legend tells how Shen Nong tasted different herbs and plants to help the earliest human beings live longer and healthier lives.

According to the legend, ancient people lived with the threat of diseases from rampant natural disasters. Shen Nong, an ancestor of the Chinese people, thrust himself forward to taste herbs and water so people could discover what was safe to eat. In risking his own life, he helped others find ways to protect theirs. Later, Xuanyuan, also known as Huang Di (the Yellow Emperor), invented characters, tones, boats, vehicles, arithmetic and silk production. He also made a great contribution to ancient Chinese medicine.

The Yellow Emperor had two medical officials in his court. One was Qi Bo who was good at prescribing medicine. The other was Lei Gong who was an expert acupuncturist and moxibustionist. Often, the three of them got together to discuss the relationship between human beings and the natural world. They studied the cycle of nature, analyzed the characteristics of *yin* (representing feminine and negative) and *yang* (representing masculine and positive), researched birth, senility, illness and death, and discovered medicines and new ways of treating diseases. Later, they wrote the great medical classic, *The Yellow Emperor's Canon of Internal Medicine*, in the form of dialogue. The book is now well-known throughout the world.

These stories show how traditional Chinese medicine (TCM) was created by legendary heroes like Shen Nong and Xuanyuan. They tell us that from the beginning Chinese medicine valued people more than anything else. It is a benevolent skill.

Sun Simiao, a famous doctor who lived in the Tang Dynasty (618-907), said, "Human life is more important than gold. Saving it means excellent ethics." He specified three principles:

First, a doctor must be kind to patients and indifferent to his own fame or wealth. He must be dedicated to developing his medical skills, and consider it his duty to save the dying and rescue the wounded. He must even be ready to risk his life to save patients from danger, but should never be bent on asking for money by taking advantage of his medical skills.

Second, a doctor must treat all people with equal kindness. He must eagerly offer genuine help to all patients, rich or poor, old or young, male or female, friends or enemies, Chinese or foreign. He should not despise the poor and curry favor with the rich.

And third, a doctor must brave fatigue and hardship to rescue his patients when they need him badly. A doctor should not think more about fame and profits than the patient he is treating. He must take the patients' pain as his own and, when needed, go to save patients without delay, not caring about time, working conditions or his own hunger, thirst or fatigue.

Only the doctor who disciplined himself with these principles was a real one, thought Sun. If he failed to do so, he could become a dangerous enemy.

These three principles summarize the experience of good ancient doctors. Here is a story about a doctor from the Three Kingdoms Period (220-280).

Dong Feng lived in Lushan Mountain, Jiangxi Province. He never refused patients no matter how ill they were. And he never received money. Instead, he asked his recovered patients to thank him by planting apricot trees in his garden. Patients who had been slightly ill were asked to plant one tree, and seriously ill patients were asked to plant five. A few years later, Dong's garden became a forest of apricot trees. In early spring, the flowers blossomed. And in summer, the trees were overburdened with fruit. Every year when the apricots were ripe, Dong put a container in his granary and a notice on the wall, saying those who wanted to buy apricots may exchange grain for them. One container of apricots was worth one container of grain. People were trusted to do the exchange themselves without telling him. Dong kept a small quantity of grain for himself and gave the rest to poor orphans, widows, childless elderly and travelers in a hurry without a hotel to eat at.

People spoke highly of Dong's manners. *Xinglin*, meaning "forest of apricot trees", became a word representing doctors. People used words such as "warmth of spring in *xinglin*" and "prestige in *xinglin*" to express their gratitude to doctors. Doctors also encouraged themselves with the word. Some hospitals, pharmacies and medical institutions were even named as "Xinglin Hospital", "Xinglin Hall" or "Xinglin Society".

Despite the thousands of years between them, the herb tasting Shen Nong, apricot growing Dong Feng and even the Canadian Doctor Berthune [ Norman Berthune, a Canadian doctor who helped China during the War of Resistance Against Japan in 1937-1945 -Ed.] are linked by their belief that "human beings are the most valuable thing in the world". This is the aim of Chinese medicine and the highest ethical principle of all good Chinese doctors.

神农尝药图　古代佚名氏画
Shen Nong tastes plants.

# Chapter 2

## Science vs. Superstition

In the distant past people were consumed with fear about the mysteries of nature. They lived in a world of superstition. Their thoughts and behavior during war, social activities and life were almost totally influenced by superstition.

However, as the people's understanding of nature and the routines of human society deepened, some began to see through their supernatural beliefs. Sun Zi, an ancient Chinese military strategist in the Spring and Autumn Period (770-476 BC), strongly criticized superstition.

He had seen from countless facts that war was a struggle between strength and wisdom rather than the gods. He suggested sorcery in military activities be forbidden, and military men be required to seek information from experts because war was closely related to national security and people's lives.

In the same way, medicine is also closely related to human life and health.

Helping people overcome superstition and develop medical science, ancient medical scientists, in addition to treating diseases, involved struggling with social and human malpractices such as superstition. This was especially important when medicine first began.

*The Yellow Emperor's Canon of Internal Medicine* says, "You can never talk about medical theory with those who believe in witchcraft. And you can never talk about acupuncture with those who are tired of it."

China had not only systemized medical theories and technologies over 2,000 years ago, but also produced a great number of medical scientists with advanced thoughts. Bian Que, who lived in the late Warring States Period, was one.

Bian Que was a famous doctor of his time. He was dedicated to treating disease without regard to fame and wealth. Sima Qian, a historian living around 135 BC, wrote a special chapter about Bian Que in his great work, *Records of the Historian*. He was the first medical scientist in Chinese history to have his own biography.

Bian Que insisted on the principle of "not treating those who believe in witchcraft". He used facts to teach people to believe in science rather than superstition.

*Records of the Historian* records this story about Bian Que.

One day Bian Que was traveling through the State of Guo when he heard the crown prince of the Guo State had died half a day before, but the court officials could do nothing but pray to God to bring him back to life. So, Bian Que went to the palace to take a look. Having realized how the prince had died, Bian Que said he could make the prince live again.

A court official named Zhong Shuzi said, "I have heard that highly-qualified ancient doctors could make the dead come back to life by treating the brain and marrow. Do you have these skills?"

Bian Que said, "If you do not think I'm qualified enough, you can go to the crown prince. You will hear sounds from his ears and see his nose open. You can also feel warmth on the side of his thighs."

The official went in and told the king, who was surprised and came out, saying, "My son can be saved. If it were not for you, he would have been buried and could never return."

Bian Que told the king that the crown prince was suffering from a cadaverous coma and was not really dead.

With the king's permission, Bian Que used acupuncture on certain acupoints on the prince, who gradually woke. Then, Bian Que applied a hot compress to the patient's chest. At last the prince slowly sat up.

For 20-odd days, Bian Que massaged the prince and concocted medicines for all his conditions.

Everyone praised Bian Que for restoring the dead

to life. But Bian Que told them, "I cannot make the dead come to. This man was alive. I only helped him to recover."

Bian Que's scientific viewpoint and modest attitude were a forceful attack on superstition.

Zhang Zhongjing, a renowned doctor in the late Han Dynasty (221 BC-220), further developed Bian Que's thoughts. He noted that many people showed little concern for medicine and the art of healing even though it would benefit their health. Instead, they sought wealth and power, hoping to become rich or noble one day. They did not value themselves or others. They only woke up when they became seriously ill. But, because they knew little about medicine, they believed in sorcery and entrusted their health and lives to it. "What do fame, power and wealth mean to them when they are dead?" asked Zhang.

Zhang's words point out the mistake made by people who care for wealth and power while neglecting their health, and illustrate the fate of those who believe in sorcery. These viewpoints are still worth highlighting today.

Even if we are not ill, it is good to understand something about medicine to help us take care of our own health and better cooperate with doctors when we are really ill.

黄帝内经
*The Yellow Emperor's Canon of Internal Medicine*.

华陀施腹部手术
Hua Tuo operates on a patient's abdomen.

## Chapter 3

# Military Preparedness and Preventative Medicine

As early as 2,000 years ago, Chinese doctors advised people to treat disease the same as war, and to practice preventative medicine. *The Yellow Emperor's Canon of Internal Medicine* notes, "Sages prevent illness before it occurs and take precautions against turmoil in times of peace. Taking medicine only when one gets ill and resorting to control when turmoil occurs are much like waiting to dig a well when feeling thirsty and forging armaments during a battle. Isn't it too late?" Its admonitions clearly illustrate the importance of preparedness and prevention.

China is a country with both an ancient civilization and a long history of war. *The Book of Changes*, written over 3,000 years ago, notes, "Remain vigilant of danger in times of peace, be cognizant of death while alive and remember the chaos of war in periods of stability." The warnings simply summarize the experience of history.

The following short account tells of how the ignorance of war destroyed a kingdom.

Duke Yi of the State of Wei was so fond of cranes that he appointed them to official salaried positions, while at the same time ignoring national defense. Soldiers of the Wei State, who considered Duke Yi a fatuous and self-indulgent ruler, were reluctant to fight during a later invasion by the State of Di. Their reluctance in turn led to the fall of the State of Wei.

Situations of this type were nothing new to history.

Sun Zi pointed out, "You must not count on the enemy not coming, but always be ready for him. And you must not count on the enemy not attacking, but make yourself so strong that you are invincible."

While the destruction of a state due to the ignorance of war is quite different from death caused by a curable disease, they nonetheless teach the same lesson. Therefore, stress should be focused on prevention and treating illnesses in the earliest stages. Prevention and timely treatment are strategically important.

Preventing disease is like deterring an enemy. The latter relies on strength, while the former on good health. In the words of one ancient book, "One should value life while healthy, be prepared for unforeseen disaster and prevent disease before it strikes."

Preventing war depends on knowing both the enemy's and one's own capabilities, while the prevention of disease centers on determining the cause. Various conditions contribute to the outbreak of disease, with the traditional Chinese medicine listing the following three main categories:

1) External factors — wind, cold, heat, humidity, dryness and fire. Disease often occurs when people are physiologically unprepared for sudden changes of natural conditions.

2) Internal, or emotional, factors — joy, anger, melancholy, sullenness, sorrow, fear and shock. While normal emotional changes will not cause disease, sudden, strong or long-lasting stimulation will quite often surpass human endurance and cause illness.

3) Various combinations of the first two categories. For example, dietary imbalances, excessive sexual intercourse, uneven allocation of work and improper care for injuries all influence physiological functions and induce disease.

Generally speaking, however, there are two main aspects of prevention — internal and external. According to one ancient book, "Carefully avoiding unhealthy environmental influences and keeping a healthy mentality help prevent disease."

For the first aspect, the book advises that people should abide by the routines of natural change and care for themselves according to the characteristics of the four seasons. There are numerous old sayings in this regard, including "wear more in spring, but less in autumn"; "keep the room and bed warm in winter and clean in summer"; and "sleep facing east in spring and summer, but west in autumn and winter".

It also cautions people to remain calm emotionally, maintain a moderate diet, avoid overwork and refrain from excessive sexual intercourse to protect their physiological functions. Ancient Chinese people practiced efficient methods for maintaining good health, including spiritual self-cultivation, *qigong* and sports.

Despite all of the advice on how to take care of oneself, doctors still make full use of medicines to prevent and treat diseases. Chinese history is filled with countless medicines and treatments for preventing disease. However, the prevention of smallpox, one of the most outstanding achievements, can be considered a great contribution to mankind.

Smallpox, a highly infectious disease, was introduced to China around the 2nd century and very quickly spread from the south to the north. Qing Dynasty (1644-1911) emperors worried that the spread of smallpox would jeopardize the health of their successors.

Chinese people began using smallpox scabs to prevent the disease as early as in the 8th century. The method to provide immunity involved blowing crushed pieces of scabs in the nose of a healthy person or otherwise vaccinating people with processed particles of scabs from smallpox sufferers. This method spread outside China in the 17th century. The British medical scientist

Edward Jenner learned from the Chinese method and invented smallpox pustule in 1796.

On October 26, 1979, the World Health Organization in Kenya announced the worldwide eradication of smallpox. Ancient Chinese people deserve greater credit for the achievement than does the British doctor.

Facts in recent decades prove that other infectious diseases, including typhus, relapsing fever, kala-azar, polio and snail fever, can also be prevented or controlled to certain extents.

Chinese doctors have researched and developed a series of herbal medicines and medical devices for the treatment of cancer, heart and cerebral diseases and hypertension. For example, innovative magnetic shoes offer a tranquilizing effect that eases hypertension. Yang Shangshan, a doctor in the 7th century, proposed that wearing shoes with magnetite could help prevent and cure diseases.

TCM offers various methods for preventing disease and protecting health. Most methods are quite simple and can be easily carried out. For example, drinking wild chrysanthemum tea in spring helps refresh the head and brighten the eyes. Drinking soup brewed with boiled talcum and licorice root helps ease internal heat and prevent sunstroke. Drinking a decoction of brown sugar and ginger helps prevent colds when one gets wet. A soup mixture of boiled lotus root and crystal sugar is good for coughs resulting from dry air in autumn, as well as for people with blood in their phlegm. Mutton creates internal heat and thus should be consumed in winter rather than in summer. Leaves of purple perilla and ginger should be used as spices to detoxicate sea food. Hawthorn boiled with honey helps improve the functions of intestines and softens blood vessels. Soup of boiled Chinese angelica, ginger and mutton is an effective treatment for women suffering from uterinary bleeding due to weakness and cold factors.

Ancient people realized that "medicines and food come from the same sources". They also knew that mixing them properly would help prevent disease and keep people healthy.

Prevention is therefore quite different from treatment. Preventing an illness is much easier and indeed much more effective than treating the same.

However, to a certain extent at least, preventing disease is more difficult than preventing enemy incursions because the latter can be easily observed, while the former is hard to detect. Prevention requires both patience and care.

宋代斗茶图
Drinking tea 1,500 years ago.

## Chapter 4

## Prepare and Prevent

Ancient people, even before Sun Zi, had said, "Morals are formed in small matters. And so are disasters." They thought that small matters were the beginning of both good luck and disaster; hence, they should not be ignored, because "it is easier to prevent small matters than to get out of a disaster".

This is a summary of historical lessons. Emperor Xuanzong of the Tang Dynasty (618-907) ignored the maxim, leading to the turmoil motivated by two court officials, An Lushan and Shi Siming.

Emperor Xuanzong, paying no attention to suggestions that An was rebellious, was so enchanted by his sweet words that he put him in charge of a force of 150,000 soldiers. When An and Shi started a rebellion in Fanyang in 755, the Emperor was preoccupied with sweet singing and graceful dancing in his capital Luoyang and showed no vigilance. The advancing rebels were able to conquer Luoyang easily. It took eight years of fighting before the rebellion was finally put down, by which time the Tang Dynasty was almost drained of its strength.

This lesson is worth learning in many respects.

Doctors of TCM long ago realized that illness starts from changes in the superficies. If it is not treated in time, it will become serious, making a cure more difficult. For example, most people think that a cold is not a serious illness and they will recover without medicines within a few days. They don't realize that a "simple" cold can develop into such serious diseases as pneumonia, tracheitis, nephritis and rheumatism, as well as leading to cardiac problems.

The same situation prevails in splanchnopathy (visceral problems). For example, liver disease may influence the spleen, and diseases of the latter may influence other viscera. That's why *The Yellow Emperor's Canon of Internal Medicine* points out, "A good doctor first treats the ailment in the skin, then the illness in the muscles, then the channels, and then the six *fu* organs[*], before reaching the five *zang* organs[**]. If the *zang* organs have to be treated, the opportunities for the patient to live or die are 50-50." This proposition demonstrates, in addition to prevention, early treat-

---

[*] Six *fu* organs, also known as six hollow viscera, refer to the gallbladder, stomach, small intestine, large intestine, bladder and triple energizer.

[**] Five *zang* organs, also known as five solid viscera, refer to the heart, liver, spleen, lung and kidney.

ment should also be stressed so that diseases can be eliminated in an early stage or prevented from worsening.

There is a story in Chinese history about a man who refused to accept his doctor's advice so that he became incurably ill.

Bian Que once went to see the Marquis Huan, ruler of the State of Qi. He found him off color. So he told the Marquis Huan, "You are ill. The trouble is in your skin. It is not serious. But, if you do not treat it in time, it will get worse." The Marquis Huan answered coldly, "I am not ill." When Bian Que left, the Marquis Huan said, "Doctors always like to tell healthy people that they are ill to show their brilliance."

Five days later, Bian Que went to see the Marquis Huan and said to him seriously, "Your illness is now in the arteries and veins. It will become more serious if you do not have it treated quickly." But the Marquis Huan did not believe him.

Another five days passed. When Bian Que saw the Marquis Huan again, he called out, "Your illness is now in your stomach. It will be dangerous if you delay." Still, the Marquis ignored his advice.

Five days after that, Bian Que, as soon as he saw the Marquis Huan, turned around and went away without saying a word. The Marquis thought this strange and sent someone to ask Bian Que what he meant by this action. He replied, "When the illness is in the skin, application of hot towels can cure it. When the illness gets into arteries and veins, acupuncture and moxibustion can be applied. When the illness enters the stomach, there is still a chance to recover by taking herbal medicines. But once the illness spreads to the marrow, nothing can be done. The Marquis Huan is now ill to his marrow."

Five days passed. The Marquis Huan, as expected, became so ill that he could not even get up. He quickly sent people to look for Bian Que. But the doctor was nowhere to be found. Soon, the Marquis Huan died.

This story is typical and enlightening. In his *Shang Han Lun* (*Treatise on Fevers*), Han Dynasty medical scientist Zhang Zhongjing said, "When illness first attacks a human body, it is in the superficies. Vital energy and various internal organs have not been influenced yet. So, it is easily treated. When the pathogen goes deep and gets mixed with the vital energy in the body, treatment is more difficult, because suppressing the pathogen will influence vital energy and supporting the vital energy will help the pathogen. Even if the pathogen disappears gradually, vital energy has been hurt. If people pay little attention to their health when they first get ill, it will get worse and become dangerous." So, he warned people to get medical care as soon as they get ill. He also suggested that doctors offer early treatment to patients so that the illness can be killed in an early stage.

Today, medical conditions are highly developed, and the idea of precautionary health preservation is gradually becoming rooted in people's mind. There is a stress not only on early treatment, but also on early discovery and early diagnosis. Facts prove that only by doing so can diseases, including cancer, be treated, and positive results achieved.

Regular physical examination can help people find diseases, including tuberculosis, tumors and hepatitis, in time so that better curative effects can be achieved, with treatment costing less. It can also help prevent infestation.

伤寒杂病论 （木刻版）
Ancient block-printed medical books.

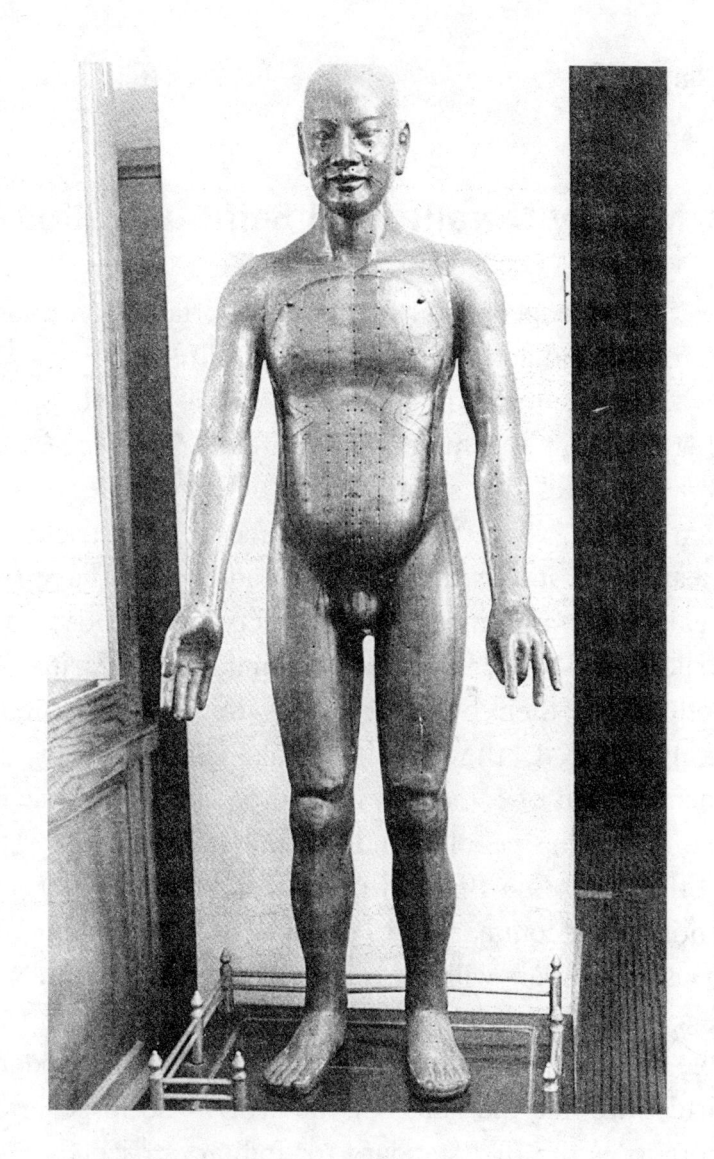

铜人模型针灸穴位图
Copper sculpture showing acupuncture points on human body.

## Chapter 5

# Military Morality and Self-Cultivation

Chinese people always lay importance on moral character and active attitudes toward life.

In the same vein, morality is also a kind of spiritual strength in the military field. Sun Zi's *The Art of War* stipulates five standards for military leaders: wisdom, faith, benevolence, braveness and strictness. Meanwhile, it says that military leaders often have five weaknesses: fighting in a death-defying attitude, they could easily be lured into a trap and killed; caring for nothing but their own lives, they might be captured; easily irritated, they might be taken in lightly by the enemy; fond of fame, they could be easily humiliated; and cherishing people blindly, they would be disturbed all the time. Sun Bin said in his *Art of War*, "It is essential for a commander to be virtuous. If he is not virtuous, his orders will not carry weight, and without orders that carry weight to direct the army, it will not be able to score victories. Therefore, the commander's virtues are like the hands of the army." So in general, morality is the first standard for military officers.

It is the same in medical science. Medical scientists

have found that self-cultivation plays an important role in maintaining health. *The Yellow Emperor's Canon of Internal Medicine* points out that self-cultivation is the key to good health and long life.

From the point of view of traditional Chinese medicine, character is the illustration of one's morality. Only by cultivating one's moral character can one be calm, modest and peaceful so that the physiological functions can be improved to adapt to external stimulation and resist disease. Without this, the spiritual and physical conditions will decline, and one will easily get ill.

*The Yellow Emperor's Canon of Internal Medicine* says, "Indulging in material desires and worrying about personal gains and losses will harm the health. If one does not make things up when getting ill, the condition will worsen, and it will be difficult to be cured."

Sun Simiao said, "While cultivating morality, you must practice ethics, live in harmony with other people and be calm in all dealings. Only when these become natural to you, can you have a moral character. Then, both internal and external diseases will find no way of attacking you, nor any troubles or disasters can influence you. This is the essential rule for preserving health."

The key to avoiding disasters and troubles is to be open-minded and magnanimous, stand aloof from worldly success and oppose fighting for fame and wealth.

There is a story of an immortal named Lu Dongbin.

When Lu heard it said that he killed people with a flying sword, he smiled and said, "The most benevolent is the Buddha. Immortals are like Buddha. How can they kill people? I do have three swords. But I use them differently from what people say. I use one to kill greed, one to kill carnal desire, and the last to kill vexations."

Of course, neither military nor medical sciences simply stress nihilism and indifference to fame and fortune. Instead, both require a proper balance between work and rest, honesty and avoidance of distracting thoughts.

*The Yellow Emperor's Canon of Internal Medicine* says, "A peaceful mind makes one focused and honest, so that the body will not be affected by unhealthy factors."

Sun Simiao, respected as the "king of medicine", made a great contribution to TCM because he was good at cultivating himself both spiritually and physiologically. His long life, more than 100 years, convincingly proves that his way is right.

Obviously, the self-cultivation theory of TCM actually means an outlook on life. Only by attaining a lofty realm of thoughts, instead of being swayed by considerations of gain and loss, and being kind and magnanimous, can people live long.

## Chapter 6

# Terrain and Topography

People are closely related with nature. It is interesting that people's choice of their living environment is somewhat in line with certain military principles.

In *The Art of War*, Sun Zi said, "A maneuvering army prefers high, dry ground to the low and wet; it prizes the sunny side and shuns the shady, so that food and water would be readily available and remain in ample supply, and men and horses may rest and restore their strength and be free of diseases. These conditions will guarantee victory."

These words not only define the basic requirement for stationing an army, but also outline the principles of the environmental preservation of health.

A health expert living about 500 years ago said people's houses should be in "high and dry places" so as to avoid peril, because the air there was fresh while that in low and wet places was dirty.

Usually, an adult takes in about 15 cubic meters of air every day. Fresh air, rich in oxygen, nitrogen and anions, is one of the essential substances that maintain the human metabolism. By comparison, dirty air brings serious harm to human health. For example, respiratory system diseases, lung cancer, stomach cancer and myocardial infarction are all related to air pollution.

So, living in high and dry places can help prevent people from the harmful effects of dampness.

No matter at work or home, people like to have sunshine in their rooms. Sunshine not only kills bacteria, helps resist illness, cleans the air and increases temperature, but also influences people's mood. Bright sunshine lifts peope's spirits.

When building houses, people always pay great attention to lighting. Though they plant trees in front of and behind their houses, they normally don't let the trees shade the doors and windows. Ancient people thought that a house good for health should "have open areas on the two sides so as to accept sunshine and have a big yard in front of the rooms so as to possess an open view and ease of mind... Trees are also needed. But they have to be artistically spaced. Otherwise, too many trees will reduce sunshine and increase gloom."

Sun Simiao's requirements for his residence were: with a hill at the back and water in the front, good climate, rich soil and clear drinking water. This reflects much the same idea as Sun Zi. Living in such an environment, Sun Simiao lived 102 years.

Water is the source of all life and a necessary substance for human beings. Where there is water, there are luxuriant plants and beautiful scenes good for human health. So, residences should be near water. Meanwhile, it should be convenient in communications. An ancient book says that people should live in a big house, have fertile farmland, plant bamboo and trees around, have an open yard in the front and an orchard behind, and have a boat or vehicles for long-distance travel.

Ancient people who enjoyed a long life all lived in the sort of environment described above. And that's why Buddhists and Taoists who wished to live long built their temples in beautiful mountains or on the banks of rivers. That's also why the emperors in Chinese history had temporary dwelling palaces outside the capital built in places where there were hills and rivers and a mild climate.

Yet, people usually cannot choose their living environment entirely according to their own will. It is especially so in modern society. For example, it is difficult for urban residents to live in mountain villages. But, people can reform and beautify their environment according to local conditions.

You Chaoshi, a legendary hero, set an example in this respect.

According to an ancient history book, in remote ages people were too weak to survive among fierce animals. A man came out boldly and made houses with wood to prevent people from being attacked by animals. People were so happy they made the man their king and called him You Chaoshi (the man who invented the house). Though the original purpose of the legendary hero was only to prevent people from the attack of animals and bad weather, his actions actually marked the beginning of the process of using the environment to keep good health.

Now, civilizations are highly developed. People are more capable and have better conditions to reform and beautify their living environment.

On the one hand, they can grow trees, flowers and grass in yards or on balconies. If conditions permit, they can build a rockery and man-made fountains, raise fish and birds, making a small garden. This helps purify the air and alleviate noise. Meanwhile, it brings fun to life. Those who do not have a yard or balcony can grow some green plants in basins in their rooms. This is also good for the eyes. In cold winter, green plants in rooms make people feel warm.

On the other hand, people should have their windows and doors open to the south and the north. Opening them often helps create sufficient light and circulate the air. Clean the houses inside and outside to prevent pollution.

Only by doing so can diseases be prevented, and people live long.

# Chapter 7

## More Is Not Always Better

Food is important to human beings because it is one of the basic conditions for maintaining life. But improper diet may cause diseases and even make people die young.

Xu Dachun, 1693-1772, illustrated the danger of improper diet by taking deploying troops as an example. "People who are fond of victory in battles surely will suffer disaster. It is the same with those who are fond of good clothes and food."

Improper diet refers to drinking or eating too much, too fine or too rich food, etc..

With the improvement of living standards in the modern society, people find it easy to drink and eat too much without thinking that they may easily get ill as a result. TCM holds that fried, roasted, stewed and boiled meat and grains, if eaten in excess, may cause internal heat so that diseases such as noxious heat, scabies, phlegm-heat, carbuncles and swelling may occur. This is in line with modern medicine, which says that eating too much of these food may easily cause hyperlipenia, hypertension, adiposis, diabetes mellitus, arteriosclerosis, coronary disease and furuncle infection.

Dietary bias is also harmful. It usually causes a nutrition imbalance, metabolic disorder and decrease of the immunological function of cells so that resistance to disease will be weakened.

An ancient medical book says, "Too much salty food makes the blood creamy and the face pale. Too much bitter food makes the skin haggard and the hair withered. Too much hot food causes muscle tension. Too much sour food causes calluses and makes the lips dry. And too much sweet food makes the bones ache and the hair drop out."

What diet is good for health, then?

More than 2,000 years ago, Sun Zi pointed out, "The strength of an army does not lie in mere numbers." *The Yellow Emperor's Canon of Internal Medicine*, noting the harm of excessive eating and drinking, sponsors a "moderate diet" in living a healthy and long life.

"Moderate diet" includes the following five aspects:

1. Do not eat too much or have food that is too tasty. And do not have dietary bias. Have a diet with proper proportion of coarse grains, wheat and rice, as well as meat and vegetables. Old people have to pay special attention to this because their physiological functions are weakening. They easily get hungry, but find it

difficult to digest if they eat too much. So, their diet must be better controlled. There is an old saying, "Eat when you are hungry, and drink when thirsty. It will be better to eat less than to eat more."

2. Keep life in routine. Do not have dinner too late. In China, people usually have three meals a day because digestion of the food needs four to five hours. It will be better to have breakfast at about seven A.M., lunch at about noon and supper at about six P.M..

3. Eat less or avoid unnecessary food, as well as food that brings bad influence to health. An ancient doctor said, "Good wine, greasy meat, too spicy noodles and too hot, sweet, baked, stewed or fried food should all be avoided." Some medical books stress, "Eat less cold fruit in autumn to avoid dysentery."

4. Food should be warm and soft and be eaten slowly. The stomach needs warmth not coldness. So, no matter if it is summer or winter, food should be warm. Some food is too hard and is not easily digested. So, this kind of food should be softened, especially for the elderly. When eating, people must chew carefully and swallow slowly so as to absorb as much nutrition as possible.

Lu You, a poet living in 1125-1210, paid special attention to this when he got old. He wrote a poem, *Porridge*, saying, "Everyone in the world wants to live long. But no one realizes that long life is here for every one. I have got the right way — that is to eat porridge."

5. Eat clean and fresh food. Do not eat spoiled and smelly food, meat of animals that died of illness and fruits falling down from trees to prevent diseases coming in by the mouth. It is said that Du Fu, a famous poet who lived in 712-770, died unexpectedly because of this. A history book reads, "When Du Fu visited Leiyang, a flood took place. Bottled up, Du went hungry for more than 10 days. The local magistrate, knowing of Du's presence, ordered a boat to fetch him. The magistrate offered Du beef and wine.... Du drank too much and died at night." But people later believed that Du died of spoiled beef.

Besides a proper diet, good eating habits are also necessary.

Experiences of people who have lived long say, "Do not talk too much when eating." "Do not rush when the stomach is full." "Do not sleep when you are full." "Massaging the stomach after eating helps prevent diseases." And "Walking slowly after eating helps accomplish a long life." By doing so, food can be fully digested and absorbed to prevent diseases caused by dyspepsia.

An ancient story says, when an old man in his nineties was asked what made him live so long, he replied, "I never eat too many delicacies."

These words should be a warning for those who are too fond of delicacies.

For thousands of years, Chinese people have been paying attention to a moderate diet and eat mainly vegetarian food. On the contrary, Westerners eat mainly meat and sweet food. That's why the incidence and mortality rates of hyperlipidemia, adiposis, diebetes mellitus, hypertension, arteriosclerosis and coronary disease are higher in the West than in China.

Facts have proven that the traditional Chinese theory of maintaining good health by practicing a moderate diet is undoubtedly scientific.

Chapter 8

## The Benefits of Exercise

A newspaper carried an article on a nonagenarian with excellent vision, hearing and mental acuity. The man, 94, offered the following response to the question of how he managed to stay healthy, "Exercise keeps my body healthy, and calmness keeps my mind alert."

Ancient medical scientists contended, "Food cures better than medicine, but is inferior to exercise." Thus they encouraged people to exercise frequently.

One ancient history book points out, "Running water never gets stale, and door hinges are never moth-eaten because they are always moving. The same principles apply to the human body. The lack of exercise will obstruct the vital energy of the body."

The same is true for everything in the world.

According to Zhuge Liang, 181-234, "Untrained, 100 soldiers are unable to resist one. Well-trained, one soldier is worth 100."

Zhuge's statements stress that maintaining an army depends on drill and practice.

Exercise and practice are equally important for maintaining good health. While most people agree, many fail to heed this advice. However, the elderly man mentioned earlier is quite different from the latter group. He described his daily regimen as follows, "I get up at 5:30 A. M. and take a shower. I massage my body for about 40 minutes under the flowing water. I eat breakfast and then walk 1,000 steps. I have persisted in the daily routine for 28 years without fail."

The most important factor to many is to persist in exercise as a form of recreation. On the other hand, however, overexercise can produce adverse results as exemplified by the following story.

One highly skilled coachman in ancient times found the opportunity to exhibit his proficiency to the emperor. All present gasped in admiration, and the emperor praised the coachman by noting that even Zao Fu, a master coachman in the past, could not be mentioned in the same breath. The emperor then ordered the coachman to make one more circuit of the grounds. The overzealous coachman continued the performance, and his team of horses finally dropped dead from fatigue.

Proper exercise is an important means of maintaining good health and is especially significant for elderly people. They should proceed in an orderly, step by step manner when exercising, with slow and simple movements representing the best approach. The best results are achieved if a person feels a sense of warmth with no fatigue, and if they have a good appetite and sleep well after exercising.

Dancing is an excellent form of exercise for the elderly. People living as far back as remote antiquity used dance as a means of maintaining physical fitness. Ancients imitated the movements and sounds of animals, while they danced. According to ancient books, people danced and prayed to the gods to relieve illness. In fact, harmonious movements of body help adjust physiological functions, enabling the smooth flow of vital energy to prevent or cure diseases.

Medical books dating back well over 2,000 years were excavated in 1973 at Mawangdui, Changsha, Hunan Province. Of the findings, one colored painting of fine brushwork, measuring 50cm by 100cm, aroused great attention. The painting, which is divided into four sections with 11 people in each section, depicted a total of 44 people performing various dance movements. Illustrations of the dancers explain the disease each movement is designed to treat. Many of the movements are based on the unique movements of various animals. The painting proves that dancing was a common method used to promote physical fitness over 2,000 years ago.

Hua Tuo, a famous doctor living from 141 to 203, documented the experiences of ancient people and choreographed five sets of dance movements according to the features of tiger, deer, bear, monkey and bird.

Hua wrote, "Exercise one set of the movements when feeling unwell and continue doing so until building up a sweat and feeling comfortable. Then, your body will feel light, and you may want to eat."

Fu Yi, a writer in modern times, wrote that dancing entertains the mind and helps prevent aging.

Dancing to beautiful music provides people with a feeling of relaxation and happiness. It also helps improve both appetite and sleep.

Numerous forms of recreation and related equipment have emerged in contemporary times. Nonetheless, achieving desire results requires one to persevere regardless of the method or equipment used.

# 一曰虎

五禽动作中的虎
Imitation of tiger's action in ancient exercises.

# 熊 曰 二

熊的动作
Imitation of bear's action in ancient exercises.

# 三曰鹿

鹿的动作
Imitation of deer's action in ancient exercises.

# 四曰猿

猿的动作
Imitation of monkey's action in ancient exercises.

# 五曰鸟

鸟的动作
**Imitation of bird's action in ancient exercises.**

## Chapter 9

## Balance Work and Rest

Maintaining good physical fitness is much like maintaining good order in a country.

One ancient doctor equated the mind to an emperor, vital energy to the people and blood to court officials. The doctor preached "loving the people to ensure social stability in the country and cultivating vital energy to ensure the retention of good health."

His point of view was actually rooted in the theory of Zhuge Liang, a statesman, strategist and prime minister of the Kingdom of Shu in the Three Kingdoms Period. Noted for his talents, Zhuge became a symbol of resourcefulness and wisdom in Chinese folklore.

According to Zhuge Liang, "governing a country is much like maintaining health. Cultivation of mind and vital energy is significant to good health, and capable people are indispensable for governing a country."

However, the famed military strategist's vast knowledge failed to help him live a long life, and he died of illness at the age of 54. Sima Yi, his contemporary, described the reason for his early death, "He worked constantly, but ate sparingly. How could he live a long life?"

From the standpoint of TCM, constant overwork caused Zhuge Liang to gradually exhaust both vital energy and blood. This in turn led to rapid aging and a premature death.

In the words of one saying, "Knowing how to work requires knowing how to rest." Despite being closely integrated, work and rest are also contradictory.

According to *The Yellow Emperor's Canon of Internal Medicine*, "Overconcentration hurts the blood; oversleep hurts vital energy; standing for extended periods hurts the bones; and excessive walking hurts the muscles."

These statements simply point to surpassing one's limits of endurance. Therefore, no matter whether an intellectual or laborer, one should maintain a balance between work and rest.

Said concept is easy for people to understand, but nonetheless quite difficult to practice.

The famous doctor Xue Fuchen, who lived about 100 years ago, died at the age of 58 due to overindulgence in both work and play. Xue was not only a dedicated medical practitioner, but was also an avid chess player. In fact, he often stayed up all night to read about or play chess. His wife became so angry that one day she threw his chess set into a well. Unfortunately, Xue refused to mend his ways and died at an early age from exhaustion.

Many intellectuals throughout history, including many in modern times, have suffered from bad health due to overwork.

Du Fu, the poet, ostensibly died at an early age due to poverty. However, another major factor leading to his death was his obsessive devotion to writing poetry.

Excessive rest is also harmful because it adversely influences the normal flow of vital energy and blood and in turn induces illness.

Sun Simiao noted, "It's good to work regularly, but not to the point your capabilities decrease due to fatigue."

As an old saying goes, "Exercising the brain helps prevent aging." Generally speaking, intellectuals live longer than most people.

Statistics suggest that the average life span of more than 3,000 Chinese scientists, artists, thinkers and members of the literati, who lived between 221 BC and the late 1940s, stands at 65.18 years, 30 years more than that of ordinary people.

An analysis by foreign scholars of Europeans since the 16th century shows that the average life span of inventors and scientists is 79 years. For example, Thomas Edison lived to the ripe old age of 84, Isaac Newton reached 85, and Albert Einstein 76 years.

Modern scientific research proves that brain stimulation slows the pace of aging and prevents senile dementia.

Good health requires a combination of mental work and physical training and labor.

Despite a lifetime of frustration, Lu You, the poet, lived for 85 years. Lu, who wrote a poem every two days, paid great attention to good habits, as well as the self-cultivation of the mind, attention to proper nutrition and sports. Lu was quite fond of sports and enjoyed playing ball games as a young man. He often played ball games with soldiers, while living in the suburbs and working with the military. Lu was already an elderly man when he returned to his hometown, but he nonetheless engaged in farm work to build both his body and mind, and in the interim wrote numerous pastoral poems. Lu also insisted on cleaning his house every day, considering the act to be a form of exercise.

"While obtaining a massage is burdensome, working the land promotes a long life," said Lu.

Contemporary people study hard to grasp new concepts and technology. At the same time, however, they must pay attention to their health by exercising and maintaining a balance between work and rest.

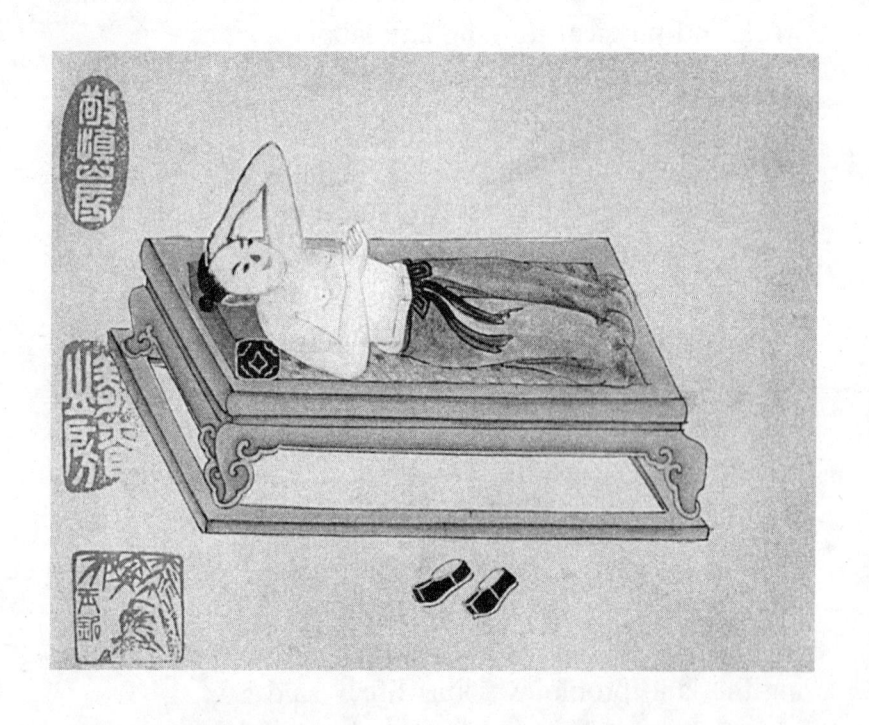

补气血按摩式　清代工笔绘画
Massage posture, nourishing *qi* and blood. Painted about 200 years ago.

治腹痛按摩式
Massage posture, treating stomachache.

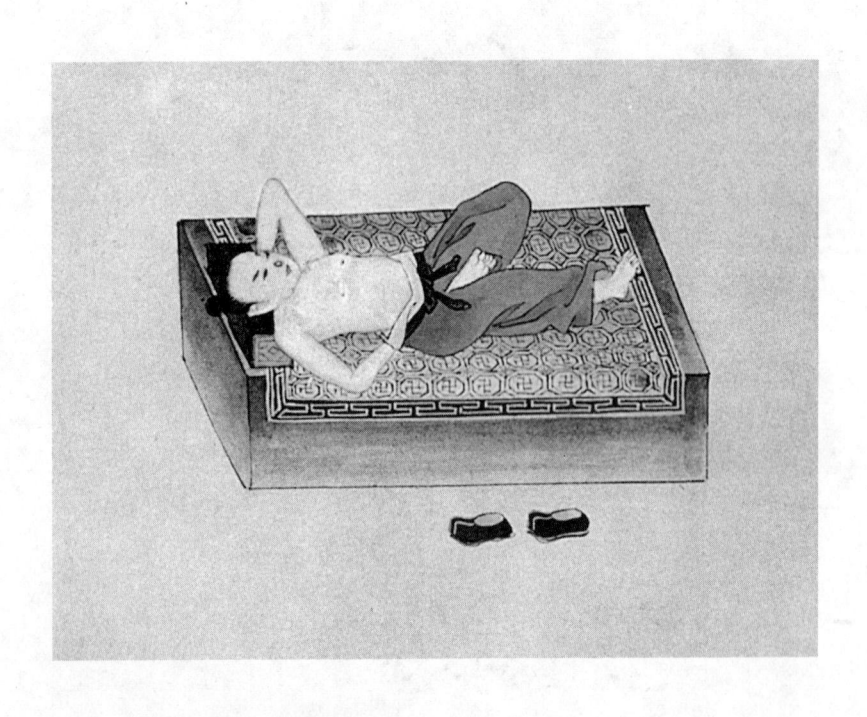

治元气虚按摩式
Massage posture, improving energy.

治晕眩按摩式
Massage posture, eliminating dizziness.

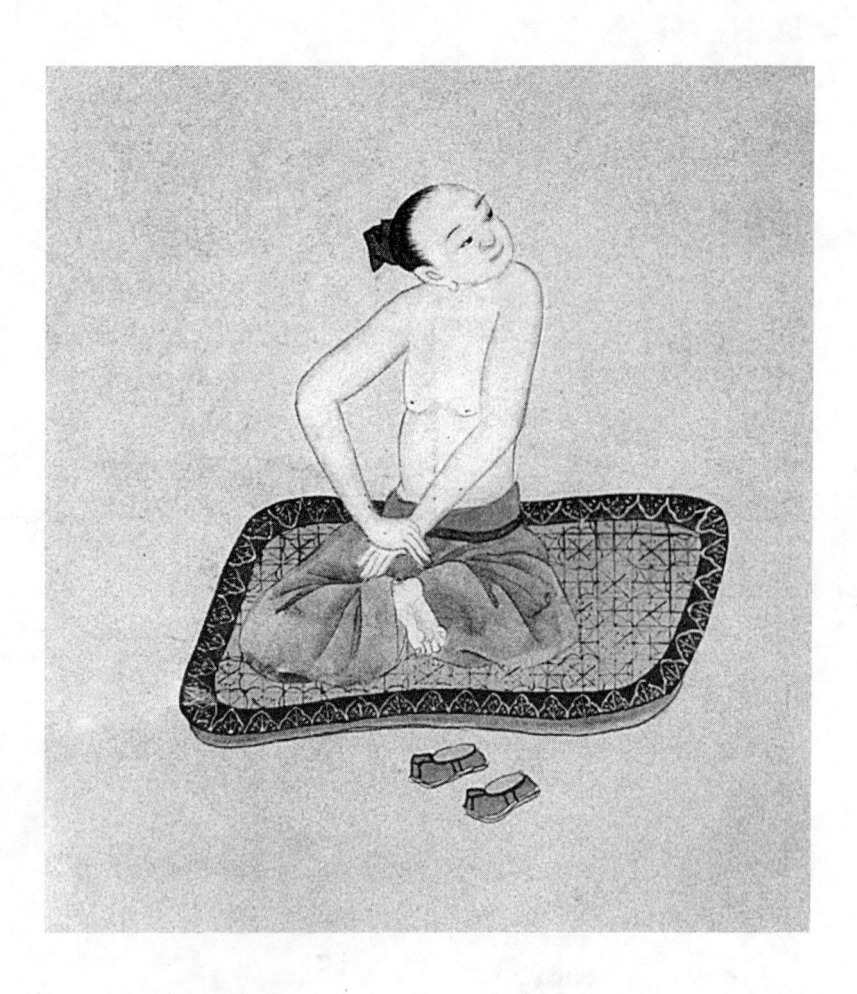

治遗精按摩式
Massage posture, treating nocturnal emission.

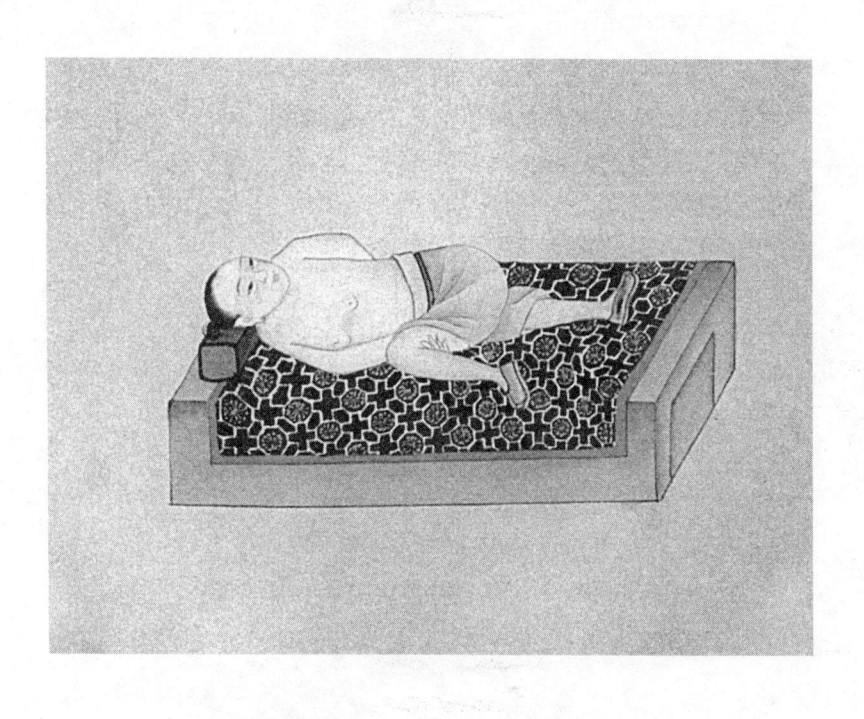

养元真按摩式
Massage posture, nourishing primordial energy.

治腰痛按摩式
Massage posture, treating backache.

炼元精按摩式

Massage posture, nourishing primordial essence.

理瘀血按摩式
Massage posture, clearing extravasated blood.

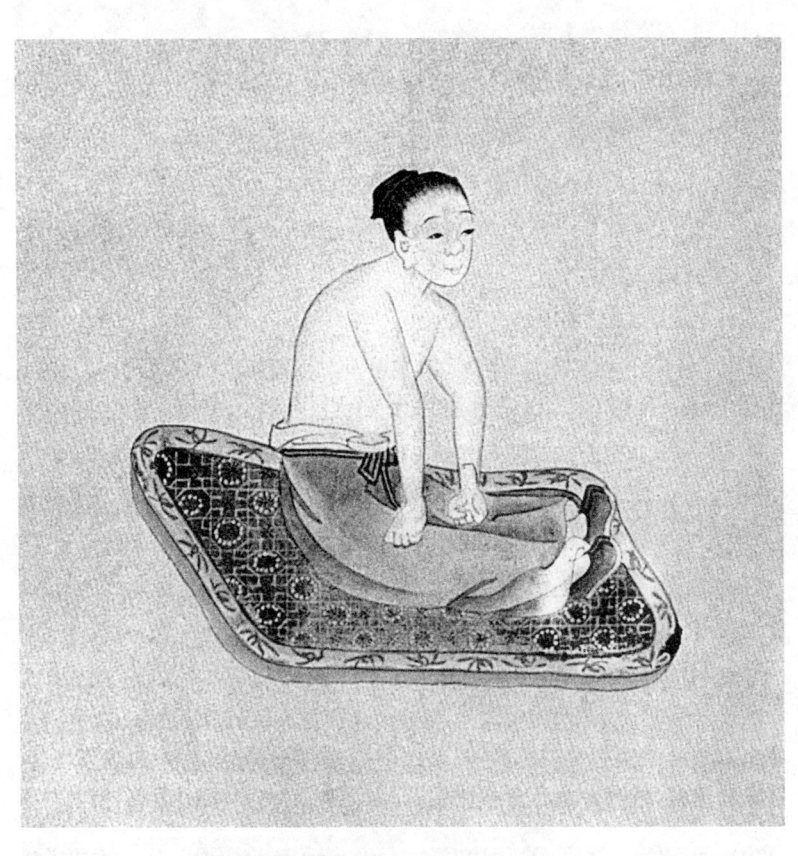

治失力按摩式
Massage posture, restoring strength.

## Chapter 10

# Anger Equals Defeat

Joy, anger, sorrow and happiness are instinctive behaviors of human beings. Normal feelings make life colorful, while abnormal feelings adversely influence human health.

Human beings exhibit various feelings. Chinese people categorize the following seven types — joy, anger, melancholy, sullenness, sorrow, fear and shock. Anger is the most difficult to control and thus creates the most harm.

Anger is difficult to control regardless of one's station in life. For example, anger not only influences the health of an ordinary man, but also disrupts the harmony of his family. On the other hand, anger not only influences the health of high-ranking leaders, but also endangers the safety of a nation and its people.

Sun Zi warned, "The sovereign should not start a war simply out of anger; the commander or general should not fight a battle simply because he is resentful."

Anger deprives people of reason and quite often simply disguises failure.

Toward the end of 219, Sun Quan, king of the Wu

Kingdom, ordered his army to launch a sne[ak attack] against Jingzhou, a city in the Shu Kingdom. Guan Yu, one of the sworn brothers of Liu [Bei, was] captured and killed. Liu Bei, king of the Shu Ki[ngdo]m, was filled with sorrow and swore to revenge the general's death. Liu personally led his army against the Wu Kingdom in 221. Liu, a leader filled with great sorrow and anger, was so eager to get revenge that he totally disregarded the enemy's strength and the surrounding terrain. As a result, the Wu Kingdom army overran his encampment, and Liu lost all of his boats and most of his soldiers. The Shu Kingdom never recovered from the failure.

Liu's anger harmed his state and his personal health as well. He collapsed both mentally and physically soon after the battle and died of illness at the age of 63.

Ancients could not help but sigh at the serious harm anger brought to people.

Lu You wrote a poem expressing his hope of overcoming anger with an attitude of standing aloof from worldly success.

According to ancient Chinese people, "Anger harms every part of the human body and must be controlled, especially when one is in agitated spirits. Otherwise danger looms, and one will have no time for regret."

According to the tenets of traditional Chinese medicine, anger mainly harms the liver. Anger causes the reverse flow of vital energy from the liver to the head, causing dizziness or headaches. An angry person's face will turn red, and in some cases anger will induce a comatose state.

Modern medical research shows a 77.3 percent incidence rate of hypertension, heart disease, ulcers and hyperthyrea in people quick to anger, while the rate is much lower to people less prone to anger. Long-term tension and worry, in fact, likely cause arythmia, heart palpitations and heart attacks. Anger causes one's blood pressure to rise. Frequent anger and repeated rises in blood pressure will in turn induce hypertension. Long-term depression creates ulcers, with poor spiritual stimulation causing diabetes mellitus or hyperthyrea.

Living a long and healthy life requires people to pay close attention to self-control and avoid moodiness.

Ancient people sought peace of mind and avoided distracting thoughts, such as honor or disgrace, promotion or dismissal, and gains or losses. The approach closely resembled the Taoist concept of letting things take their own course.

Obviously, unlike their ancient counterparts, contemporary people cannot remain aloof of the world about them. Nonetheless, retaining good health requires them to engage in daily self-cultivation.

The best approach for maintaining good health is to control emotions in order to remain calm and happy at all times.

In the words of one old saying, "Anger makes people old, but smiles keep them young. A smile makes one 10 years younger, while worry brings white hair."

One would be well-advised to keep these words in mind.

# Chapter 11

## Mental Calm Promotes Health

Many Chinese idioms include references to calmness of mind, such as "calm spirit and even temper", "be free of mind and happy of heart" and "keep a pleasant and peaceful frame of mind".

Health conscious people pay great attention to cultivating a calm mind, considering it the basic factor in remaining healthy.

Ancient Chinese people considered calmness the root of a long life.

Tao Hongjing, a medical scientist who lived in the Northern and Southern Dynasties (420-581) said, "People who maintain a calm mind live a much longer life than those with a quick-temper."

Remaining calm not only helps preserve one's health, but also cultivates one's character. Authors of ancient classics joined medical scientists in stressing the need to remain calm. Sun Zi's *The Art of War* notes, "It is the responsibility of the commander to be calm and inscrutable, to be impartial and strict in enforcing discipline."

Chinese and foreign military strategists throughout

the ages have cherished the words of Sun Zi.

One story in *The Romance of the Three Kingdoms* illustrates Zhuge Liang's calmness in defeating the enemy.

Ma Su, a general in the Shu Kingdom, lost Jieting due to overconfidence and personal prejudice. Wei Kingdom General Sima Yi's army of 150,000 men pressed on toward Xicheng, the base of Zhuge Liang, a group of civil officials and 2,500 soldiers. The people, including all officials, turned pale with fright at having no time to call on reinforcements. However, the calm and steady Zhuge Liang simply told the local residents to hide. He then opened the city gate and ordered a group of soldiers to dress as local residents and clean the streets near the gate. Zhuge Liang then climbed atop the city wall and began playing the *qin* (a seven-stringed plucked instrument) in a carefree and leisurely manner. Sima Yi viewed the unlikely scene as his army grew near the city gate. Believing there must be a massive force lying in wait for them in the city, Sima Yi ordered his forces to retreat.

This particular story serves as an excellent illustration of Sun Zi's words.

No matter whether a general, civil official or an ordinary person, accomplishing a calm mind requires one to pay great attention to the following aspects:

1. Cultivating fine personal qualities.

Traditional Chinese medicine promotes the cultivation of both health and personality. Medical practitioners stress that morality and kind-heartedness, rather than seeking private interests and personal gains, are beneficial to the realm of thought and the harmony of vital energy and blood. These beneficial traits in turn promote normal physiological activity and provide one with energy and strength.

Confucius, 551-479 BC, founder of Confucianism, said in *The Doctrine of the Mean*, "People with lofty ideals and morality can live long lives."

According to Sun Simiao's point of view, when one cultivated a moral character, he could live in harmony with other people and deal with everything calmly. Then, no diseases could have any way of attacking him, nor could any troubles or disasters influence him.

2. Being indifferent to fame and gain and practicing nihilism.

People, living in a complicated world, are inevitably troubled by and face emotional changes resulting from chaotic situations, such as failed love affairs and frustrating work. Related psychological reactions are quite normal. However, prolonged stress creates abnormal moodiness characterized by worry, sighing or sudden outbursts. Said factors in turn induce psychological dysfunction.

According to *The Yellow Emperor's Canon of In-*

ternal *Medicine*, "be indifferent to fame and gain and practice nihilism. Do not cherish vain hope and desire. Vital energy will flow smoothly, and you have a calm, healthy mind. How can illness occur?"

Legendary hero Peng Zu is said to have lived for 800 years. One ancient book revealed Peng Zu's secret to long life. According to the work, "do not desire women or song. Do not dwell on victory or failure. Do not worry about setbacks. Do not care about gains and losses. And do not seek glory. So doing will free one of a troubled mind."

Remaining calm and content is beneficial to adjusting emotions and promotes the smooth flow of vital energy and the normal operation of human organs.

3. Retaining self-control.

Retaining a calm mind requires one to adhere to the following two principles.

Maintaining control of one's own life. One should depend on oneself to properly deal with difficult people and situations. Life for most people is far from easy. However, one should learn to practice self-control and remain calm when dealing with frustrations.

Calmness depends on one's mind instead of the environment. A calm mind enables one to adopt a nonchalant attitude toward a noisy and chaotic environment. On the other hand, however, one will never attain a peaceful mind without calmness, regardless of being surrounded by a quiet and beautiful environment.

Medical scientists have long explored ways to retain calmness in terms of both the mind and environment.

One ancient book stipulates five types of behavior and ten enjoyable acts to adjust emotions.

It lists the five behaviors as sitting quietly, reading, appreciating flowers, grass and trees, chatting with friends and teaching.

The ten enjoyments are reading, practicing calligraphy, sitting quietly, chatting with friends, drinking small amounts of liquor, watering flowers and planting bamboo or trees, listening to music and appreciating birds, burning incense and drinking tea, climbing city walls or mountains and playing chess.

One intoxicated with traditional cultural activities quite naturally becomes indifferent to fame and gain and in turn remains calm in all situations.

Possessing the quality of remaining calm even when faced with danger requires that one have the courage to face reality and harbor a love for life

According to Su Zi, "A wise general must be good at evaluating both advantages and disadvantages. By considering the advantages when faced up with difficulties, he will be able to accomplish great tasks. By considering disadvantages when everything proceeds smoothly, he will be able to avoid possible disasters."

While preserving health nowadays, doing what Sun Zi said is undoubtedly beneficial to self-adjustment and the stabilization of mood. Furthermore, retaining a calm mind and living contentedly help people have good health and live long lives.

# Chapter 12

## Regularity Is Significant

In the words of one old saying, "Armies are to be maintained over the course of long years, but to be used in the nick of time."

Sun Zi's *The Art of War*, a book on military strategy, proffers various suggestions on maintaining an army, with many also suitable for preserving health.

For example, Sun Zi warned, "Do not let your forces get worn down, boost their morale and save their strength." Trained soldiers should lead a regular life free of exhaustion and slackness.

The most difficult aspect in training an army is to cultivate the consciousness of soldiers in order to ensure they are disciplined and maintain required standards for appearance and bearing no matter in times of peace or war.

A regular life is not only a prerequisite for maintaining a powerful army, but is also highly beneficial to ensuring one's health.

*The Yellow Emperor's Canon of Internal Medicine* includes a comparison between people who lived 100 years and those who lived a short life. The analysis shows that one of the major reasons for the difference centers on whether or not one leads a regular life.

In TCM, the term "regular life" covers aspects such as work, rest, manner of speech and behavior, and taste in clothing.

Historically speaking, regularity leads to a long life. Ye Shengtao, a contemporary writer and educator, lived for 94 years. One of his secrets of longevity was leading a regular life. He once said, "I have a habit of eating breakfast at 7:30 A.M., lunch at noon and dinner at 6:00 P.M.. I sleep a lot, go to bed at 8:30 in the evening and arise at 6:30 in the morning." Ye also insisted on reading in the morning, and bathing in the afternoon, even on weekends when most other people were enjoying themselves. Leading a regular life brought good health to Ye, and he seldom suffered from illness.

What then is the appropriate routine for the human body?

TCM theories consider human beings to be creations of nature. The Heavens and the Earth are the source of life, and all human physiological and life routines are closely related with changes in nature. For example, it is warm in spring, hot in summer, cool in autumn and cold in winter. Correspondingly, *yang* (positive) elements appear in spring and grow in summer, while *yin* (negative) elements emerge in autumn and grow in winter. Each day, the human body experiences changes in *yin* and *yang* during night and day. Generally speaking, *yang* prevails, and the human body is filled with vigor from daybreak (sunrise) till noon, a period in which people are energetic and efficient. However, *yin* gains strength from noon to dusk (sunset), and although still active, people nonetheless begin to tire. *Yin* controls the body at night, and people need sleep.

*The Art of War* compares changes in the human body to the morale of an army during different periods of a conflict. The work espouses the principles that "he who is skilled in war avoids the enemy when his spirit is high, and strikes when his spirit drains," and "the high morale of soldiers at the beginning of a campaign gradually fades and eventually disappears."

An interesting point well worth noting is that changes in *yin* and *yang* forces in the human body coincide with the modern theory of the biological clock, a theory that outlines regular change in the body over a 24-hour period. For example, most functions of the human body are in a state of inactivity between two and four in the morning. While people are most efficient at 10:00 A.M., hormonal changes bring tiredness at around one to two in the afternoon. Extroverts are creative at 3:00 P.M., while introverts are most often in low spirits. The effectiveness of many functions of the human body declines at 10:00 P.M., and people should go to bed.

Soldiers lead a strictly disciplined life, while doctors recommend a similar lifestyle for ordinary people. Simply speaking, "people should live a regular life."

Ancient people considered "clothing and food as the keys to good health." They contended that people should eat only what was needed and wear clean and well-fitting clothing.

The ancient philosophy on life cautioned that "one should not live in an overly large house because of excessive *yang*, nor should they live in a small house because of excessive *yin*." Ancients also contended that one should walk as light as the wind, sit as steady as a bell and stand as firm as a pine.

Ancient medical practitioners also stressed sleep. They noted that sleep helped restore energy and cultivate vitality. It is especially beneficial to the elderly.

People should both retire and rise early in the spring and take an early morning walk to stretch the muscles. Elderly people should take a nap at noon. In summer, people can sleep late, but should get up early. In autumn, they should both rest and rise early. People should go to bed early in winter to preserve *yang*, but get up late to protect *yin*. Keeping said schedules maintains a balance between *yin* and *yang* and protects people from the pathogenic factors present in nature.

Hence, regularity helps prolong one's life.

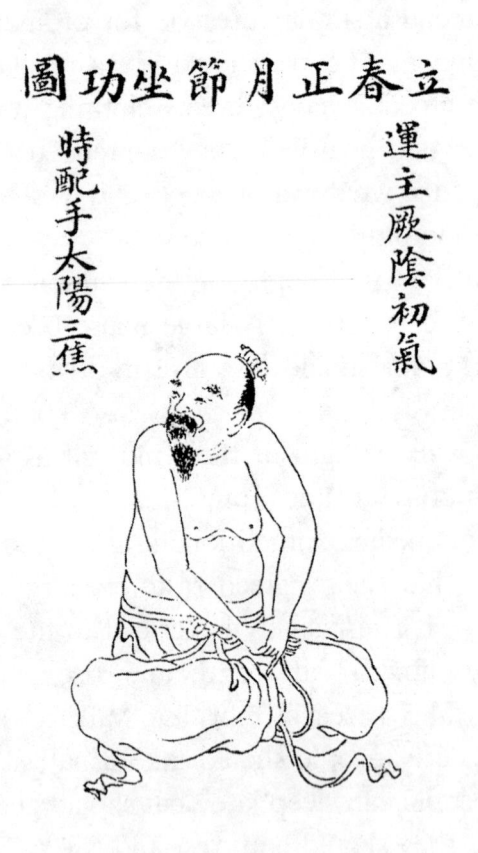

立春坐功图/北宋著名道教气功养生家陈希夷所创制。立春坐功导引，主治风气积滞及手太阳三焦经循行部位的各种病症。

Sitting Posture at the Beginning of Spring. The sitting posture, developed by Chen Xiyi, a prominent Taoist during the Northern Song Dynasty (960-1127), is said to cure stagnation of *qi* and other morbid conditions along the three-*jiao* channel of the hand-*Taiyang*.

立夏四月節坐功圖

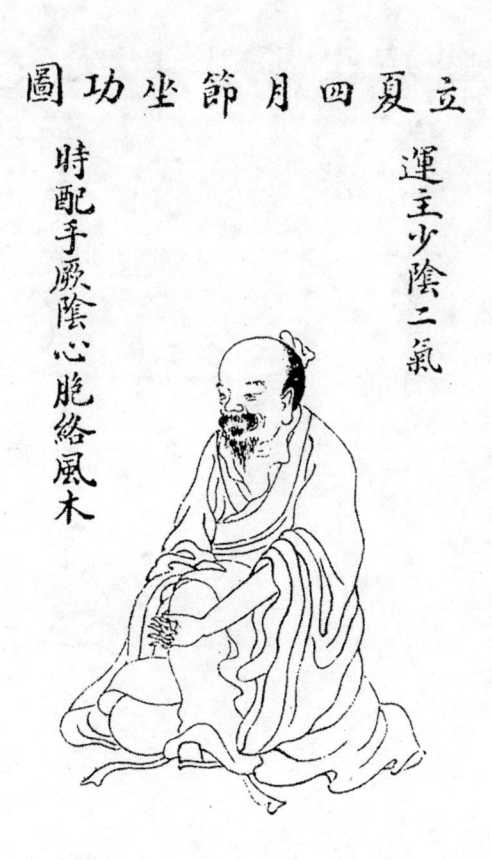

運主少陰二氣

時配手厥陰心胞絡風木

立夏坐功圖/主治风湿留滞经络肿病、臂肘挛急、腑肿、手心热、喜笑不休等。
Sitting Posture at the Beginning of Summer, designed to treat wind-dampness, stagnation of *qi*, channel and collateral dropsy, heat in the palm and unstoppable laughing.

立秋七月節坐功圖

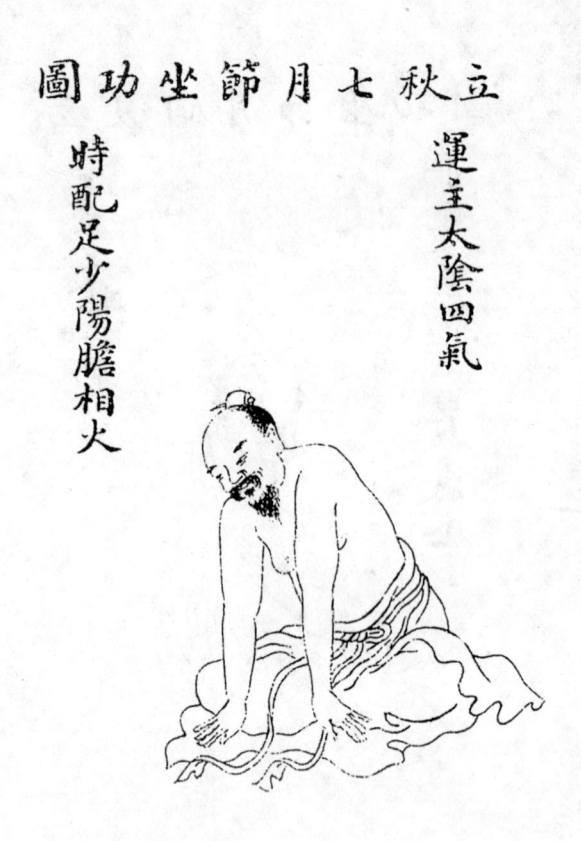

運主太陰四氣

時配足少陽膽相火

立秋坐功圖/能补益虚损,去腰积气、口苦、太息、胸胁病不能反侧等病。

Sitting Posture at the Beginning of Autumn, said to be capable of invigorating body resistance, promoting the flow of *qi*, removing a bitter taste in the mouth and sighing and eliminating pains in the chest and costae.

立冬十月節坐功圖

運主陽明五氣

時配足厥陰肝風木

立冬坐功圖/主治虛勞、邪毒、腰病、咽干、面垢脫色、头痛、耳闭、目赤肿痛等肝病症。

Sitting Posture at the Beginning of Winter, said to be capable of curing diseases related to fatigue, pathogenic toxics, lumbago, a dry throat, headache, deafness, conjunctive congestion and the liver.

# Chapter 13

## Excessive Maintenance Is Bad

A highly conspicuous characteristic of traditional Chinese philosophy centers on the golden mean — knowing when and where to stop.

Military thoughts oppose wanton engagement and aggression. The more battles a state wins, the greater likelihood of an early decline.

Enlightened ancient people realized that proper limits should be maintained in ensuring good health. They also realized that any approach to excess would lead to negative results. Ancient people understood that "color can both entertain and hurt the eyes, with sound having the same effect on the ears. What one sees and hears can entertain, but can also hurt the heart if carried to excess." Appropriate levels provide the best results.

Constantly improving living standards have been accompanied by the increasing popularity of television, with colorful TV programs bringing great enjoyment to the people. Nonetheless, facts suggest that watching TV for extended periods harms one's health.

The uneven distribution of rays and brightness levels on television screens create constant tension on the eye muscles. A viewer's eye muscles tire from constant adjustments to the distance from the TV set or from extended periods of looking at the screen. Symptoms of eye strain include dryness, a burning sensation, scratchiness, swelling and pain. The symptoms quite often lead to blurred vision and can induce near-sightedness over the long term.

People who consider it impossible to stop watching TV can nonetheless avoid excessive viewing.

Proper nutrition is one of the most important factors in maintaining life. Rationally balanced nutrition is especially significant for living a long life.

According to *The Yellow Emperor's Canon of Internal Medicine*, "A moderate diet and a regular life promotes long life."

There is also an old saying, "Avoiding both hunger and overeating is necessary for good health."

Facts prove that excessive food and drink are among the key factors inducing disease, a point clearly illustrated in Chapter 7.

Music makes life colorful and is considered the spiritual food for human beings. For example, classical music is inspiring, while light music is amusing and makes people feel both relaxed and happy. Ancient people said, "One can listen and interpret a musician's meaning. Music promotes intelligence."

Music is highly beneficial to molding one's temperament and promoting vigor.

Nonetheless, some people blindly seek stimulation by indulging in loud music, while other people appreciate listening to music through earphones. Said practices inevitably cause an imbalance between the excitation and inhibition of the brain and quite often cause diseases of the central nervous system, as well as the cardiovascular system, internal organs and digestive systems. The practices can also induce high blood pressure, arythmia and blindness. Loud music also contributes to aging.

Maintaining good health requires one to be moderate in listening to music or watching television, and in both eating and drinking. *The Yellow Emperor's Canon of Internal Medicine* notes, "Disease originates from excessiveness." The words, written over 2,000 years ago, remain perfectly applicable even today.

# Chapter 14

## Sex Drive Good and Bad

The scientific research of sexuality in ancient China was different from that in the West. Instead of paying attention to reproduction and sexual enjoyment, Chinese researchers stressed health.

For thousands of years, Buddhism and Taoism had quarreled about the issue. Buddhism, sponsoring abstinence, claimed the semen was the basis of vital energy and should never be wasted. Any wasting of the semen would be harmful to health. One ancient medical text insisted, "A couple that is healthy is one in which the individuals do not sleep in the same bed. And the individuals who know how to maintain their health will not share a quilt with their mate. There is no medicine that can compare with sleeping alone."

Taoists, on the other hand, sponsored indulgence in sex. Adherents believed in the duality of nature, a balanced struggle between *yin* (the feminine) and *yang* (the masculine). The mingling of *yin* and *yang* during intercourse brought vitality and long life.

Military scientists also commented on this issue. A military text dating to some 2,000 years ago says, "Sex is a blade that cuts through bone. If the desire is not controlled, sex will destroy you."

Some medical scientists, basing their comments on research into the biological routines of the human body, gradually came to acknowledge the nature of sex and put forward opinions aimed at clarifying the issue amongst laymen. They said:

1. Sex is impossible to ban.

Confucius said, "Food and sexual desire are the natural needs of human beings."

Ge Hong, a medical scientist living more than 1,500 years ago, said, "People cannot live without *yin* and *yang*. With no intercourse between the two, people may suffer from spiritual depression. Then illness may occur, and lives will be shorter."

2. Sexual acts should not be conducted at an early age.

According to an ancient classic, *The Book of Rites*, the right age for a man to marry is 30, while a woman should marry at 20 years of age.

In ancient times, a male was considered as adult at the age of 20, while a female was seen as a woman at 15. Only adults were allowed to marry.

3. Do not have sex at an old age.

Ancient medical scientists thought that as a man reached an advanced age, his penis withered, and that if he forced himself to have intercourse, he would not be able to ejaculate, thus causing serious internal damage. Furthermore, urination would be difficult and painful.

4. Do not indulge too frequently in sex.

An ancient poem warns, "Young women are charming. But they kill men. Though they do not cut their throats, they dry a man's marrow."

Duke Ping, a ruler of the State of Jin more than 2,000 years ago, frequently indulged in sexual pleasure. Once, Zi Chan, a senior official from the State of Zheng, visited Duke Ping and laid out for the ruler a code of daily conduct. He said a daily routine should include listening to official reports in the morning, visiting local officials and offices in the afternoon, approving state policies in the evening, and having a good rest through the night. However, Duke Ping preferred dallying with the women of the court and eventually fell ill.

The ancient Chinese had a philosophical saying, "Sex is your beginning, and sex will bring your end."

The right attitude toward sex is to know and follow the routine. *The Yellow Emperor's Canon of Internal Medicine* says, sexual desire and reproduction are biological instincts of human beings. Sexual and reproductive abilities depend on the vigor of the kidney, which develops from weak to strong and then to weak again during the life span of a person.

Generally, the vigor of the female kidney peaks at the age of 14 and is evidenced by menstruation. The vigor of the male kidney peaks at 16 and is evidenced by ejaculation. At this point, sexual maturity is reached, and reproduction can take place. At about 40, the vigor of the human kidney, both male and female, begins to fade, and sexual desire diminishes. Menopause occurs among women at or around the age of 49, while the sperm count of a man diminishes at or around the age of 64 — the vigor of the kidney nearly disappears, and humans lose their reproductive abilities.

The vigor of the kidney, the basis of other viscera, decides the levels of other functions throughout the body. Therefore, TCM suggests it is unnecessary to ban sexual activity, but it is also not wise to overindulge in it. Only by controlling the balance between restraint and overindulgence can one remain healthy.

Sun Simiao said, "Between the ages of 20 and 29, couples should make love once every four days. Between the ages of 30 and 39, they should indulge in sex once every eight days. Between the ages of 40 and 49, they should have intercourse once every 16 days. Between the ages of 50 and 59, once every 20 days is the limit. And after the age of 60 they should abstain from sexual activities, or indulge no more often than once each month."

Modern medicine agrees in a sense with the teachings of Sun. But there are no fixed schedules for sexual intercourse, because each individual has a unique health condition. Frequency depends on effects. For instance, if an individual is tired the day following intercourse, then he is acting too frequently. In general, young couples may have intercourse once or twice a week. Middle-aged couples may copulate about once a week, while older people, due to a waning desire and growing inability, should only have relations once every 15 to 30 days.

Again however, there are exceptions. Middle-aged people who are in very sound health may indulge themselves as if they were young.

A survey of 260 elderly people suggests sexual relations aid in adjusting the emotions of the older couples, giving them confidence in the face of diminishing social roles. One couple, both in their 100s, said they had continued sexual relations well into their 80s. Another couple said they made love once a week although they were in their 90s.

Banning sexual desire is harmful to older people, because the functions of their sex glands will decline, thus they will age more rapidly. Appropriate sexual intercourse postpones not only the atrophy of sexual organs but also the decrepitude of the brain.

When making love, a couple should pay attention to proper methods and rhythms. Both should be in a happy mood, showing tenderness for the other and achieving orgasm.

Modern medical research suggests that harmonious sexual intercourse depends on whether a couple understands the physiology and psychology of the opposite sex. Man needs only two to six minutes from the beginning of intercourse to orgasm and ejaculation. However, it can take a woman from 10 to 30 minutes to reach orgasm. In addition, the erogenous zones of the male body are concentrated on the penis. Therefore direct intercourse is enough to stimulate an orgasm. However, the erogenous zones of the female body are located at the breasts, back, neck and thighs. Only by prolonged manipulation of these points can a woman be stimulated. Therefore, when making love, a man should control his desire and patiently and tenderly massage his partner's erogenous zones, and the woman should take an active role in ensuring her own satisfaction, allowing both man and woman to reach mutually satisfying climaxes.

Of course, it is important to emphasize health care and hygiene at all stages of sexual activity. Medical scientists throughout history have all believed diseases are closely linked to unclean sexual intercourse. So they have suggested several guidelines.

1. Do not make love when drunk.
2. Do not make love when exhausted.

3. Do not make love when angry.
4. Do not make love when the woman is menstruating, pregnant or nursing, or just after she has given birth.
5. Do not make love when ill.
6. Do not make love when weather changes drastically.
7. Do not make love late at night.
8. Do not make love after a long journey.

Sexual relations are more than just sexual intercourse, but also involve the psychological contentment of the man and woman. Men should pay more attention to their language and emotional exchanges with their partners.

# Chapter 15

## Qigong and Morale

Among the traditional ways of maintaining good health, rhythmic and guided breathing are stressed. For thousands of years, people have found the benefits of such exercises.

*Qigong* is the practice of inhaling the fresh and exhaling waste. It is difficult to estimate when the Chinese first began practicing *qigong*. The character for *qi* appears in inscriptions on bones and tortoise shells dating back over 3,000 years. *Qi* (energy of life) flows through the body. When the flow of *qi* ends, life also ends. It is shapeless and yet exists everywhere. It is considered the origin of everything. It has become an abstract concept.

During the Spring and Autumn Period (770-476 BC), philosophers and thinkers, who advocated materialism, considered *qi* the manifestation of the human spirit. When a person was energetic, he was said to have a "living *qi*". But when he was low in spirit, he was said to have a "dead *qi*".

From this point of view, Sun Zi illustrated the *qi* of an army in his *Art of War*. The high spirit of an army was called "sharp *qi*", while the low spirit was called the "lazy *qi*". This is why Sun Zi advocated resting an army and supporting morale before battle, as well as attempting to boost morale during battle.

"He who is skilled in war avoids the enemy when his spirit is high, and strikes when his spirit drains. This is how he copes with the question of morale (*qi*)," said Sun Zi.

All discussions of *qi* in *The Art of War* are referring to the morale of the army.

But, is the *qi* discussed by Sun Zi the same as the *qi* in TCM?

*Qi* was considered the origin of all life on earth. Confucius said, "Man has three commandments: when young, his blood and *qi* (energy) are not sufficient, thus he must practice restraint in sexual desires; when middle-aged, his blood and *qi* are impetuous, thus he must practice restraint in fighting; and when old, his blood and *qi* are exhausted, thus he must go easy in both gains and losses."

The blood and *qi* here are at once biological and psychological, as well as material and spiritual. They are therefore the same as the *qi* referring to the morale of the army.

How are blood and *qi* maintained?

In ancient times, the Chinese people said, "Only by staying away from sexual, violent and material desires can blood and *qi* be maintained and vitalized."

Sexual, violent and material desires are all pathogens that harm the human body. So, people should exhale waste and inhale fresh air to maintain *qi*. This is the goal of the practice of *qigong*.

There are various ways to practice *qigong*. One is simple and easy to learn. Quite simply, the practitioner inhales through the nose and exhales through the mouth. With each exhale, the practitioner pronounces *chui*, *hu*, *xu*, *he*, *xi* and *si*. Concentration is important during the exercise.

*Qigong* is not mysterious or magical. The basic routine steadies the breathing and calms the person.

Some practitioners have added a level of idealism and mysticism to the health exercises, making *qigong* a magical treatment of all illnesses.

In fact, *qi* represents neither the good nor the bad. And it cannot influence human activities and social life. People should discard the idea of explaining social problems and predicting disaster or happiness through *qi*.

此咎欲止痨嗽
如何曰宜蹲踞
以兩手按於腦
後閉息瞑目運
其氣至膀胱穴
鳴則火性歸水
而嗽自可止矣

止痨嗽导引式　清代工笔绘画
Physical and breathing exercise to treat phthisical cough, a fine brushwork of the Qing Dynasty.

欲融會正氣如
何日宜閉息實
目正坐以兩手
抱雙膝左右盡
力而默運其氣
從小便而出乃
能脫體自得仙
道耳

融会正气导引式
An exercise to integrate the positive *qi* of the body.

或問濕腫如何
曰宜屈膝坐伸
兩手攀一足盡
左右膝中力放
而復收俟四肢
汗出是運滯血
濕腫之患

治湿肿导引式
An exercise to treat damp-edema.

欲養正氣如何
曰宜穿膝坐纍
手按脛忘言忘
怒忘樂閉息默
運叩齒氣足而
止則心自正諸
欲可戒

养正气导引式
An exercise to nourish the positive *qi* of the body.

或問感氣停食如何曰宜平立退步拗頭左右顧易左右如引弓以運片時主散氣食之養也

治食积导引式
An exercise to treat indigestion.

欲煉元神如何
曰宜屏氣瞑目
穿膝坐伸兩手
上擎左右舉力
六七度叩齒嚥
液自無昏弱之
患

修炼元神导引式
**An exercise to train the brain.**

問羸弱如何曰
宜屏氣跪坐虎
視其目以兩手
托後俟氣足叩
齒嚥液能健脾
補腎

补虚导引式
An exercise to tonify deficiencies of the body.

或問氣不能舒
如何曰正立權
謹兩手擎止徐
行百步閉息叩
齒以運氣足遂
止其鬱結之患
而自釋矣

舒气导引式
An exercise to promote the flow of *qi*.

諸欲既難戒性
敢問養心如何
曰屏氣兇視以
一手托腎絶非
禮之思默運片
時能清心寡慾
而得仙道者也

养心导引式
An exercise to nourish the heart.

欲诸经却病如何曰宜反身而卧以被经体露其手足额则金木水火土位定而精神气之本固矣

诸经祛病导引式
An exercise to treat diseases associated with the channels.

或問寒熱攻伐如何曰宜穿膝坐拗頭左右顧以左手盡力托俟額汗出充散風氣寒熱自退

治寒热导引式
An exercise to treat cold-heat syndromes.

欲養血脈如何
曰宜平立徐步
以兩手左右舞
兩足左右蹈運
片時叩齒三十
六養血脈手足
痿痺不仁

养血脉导引式
An exercise to nourish blood vessels.

若問身之衰弱
如何曰宜仰臥
以兩手抱雙膝
左右盡力依法
而臥則氣充榮
而病却延年氣
自壯矣

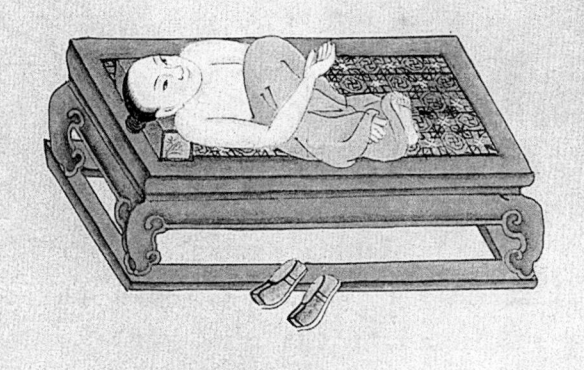

治衰弱导引式
An exercise to strengthen the body.

**Chapter** 16

# Choices Are Important

Liu Bang, a successful ruler in the Western Han Dynasty (206 BC-24 AD), said, "Misusing military officers brings about complete failure." Zhu Yuanzhang, an emperor more than 1,000 years later, said more clearly, "Encyclopedic, wise, benevolent and brave officers must be chosen to lead the battle."

During the Battle of Changping in 260 BC, the king of the State of Zhao appointed Zhao Kuo to lead his armies. Because Zhao Kuo was only good at talking about leading an army and had never actually led troops into battle, the Zhao armies were defeated, and the State of Zhao was nearly conquered.

Zhuge Liang, considered by many as the embodiment of wisdom, also made mistakes in choosing officers. He sent Ma Su, who liked to exaggerate, to guard Jieting. But Ma lost the city. Zhuge had to retreat from a battle that he should have won.

Choosing officers is important because there are differences between the good and the mediocre. This is also true when choosing doctors.

The ancient Chinese said, "In the past, mediocre

doctors killed their patients. Now, perhaps they don't kill their patients, but neither do they cure them. They simply prolong the inevitable."

In modern society, though medical sciences have made great advances, there are still those who practice medicine and yet do not have the best interests of their patients at heart. There are those who are irresponsible and incompetent.

For example, a man constantly felt chilled during early autumn. He had a fever and a cough. He visited a folk doctor, who prescribed diaphoretics. Instead of recovering, the man became weaker each day. In the winter, the folk doctor told him to drink ginseng tea. The patient lost his appetite, lacked strength and suffered from diarrhea. He soon became critically ill. His family sent him to a hospital. The doctors at the hospital quickly diagnosed his illness and with timely treatment saved his life.

During the whole treatment period, the folk doctor neither asked the patient about his illness nor analyzed the patient's conditions. He nearly killed the man.

When ill, it is better for people to steer clear of mediocre doctors. People can recover from some illnesses by relying on their own natural resistance. Although they might suffer, they are better off than if they take the advise of mediocre doctors.

The ancient Chinese said, "Patients should refrain from hoping to get well quickly but using too mild a medicine. Medicines can be both beneficial and harmful. Do not take medicines if they are not properly prescribed by a qualified doctor. You will regret not having taken medication if your condition becomes worse. And you will also regret if your condition worsens because you have taken the advice of a mediocre doctor. The latter will be even more painful."

Of course, this does not mean that people should not visit a doctor when they are ill, even if it is a mild illness.

There are many folk doctors even in modern China. Posters advertising "ancient family remedies" or "AIDS cures" are seen pasted on walls and telephone poles. An eight-year-old boy suffering from epilepsy was given an "ancient family remedy". Instead of curing his illness, the medicine made him very sick. The pain in his stomach was so severe, his family rushed him to the hospital. The experts at the hospital discovered the boy had lead poisoning.

Although mediocre doctors are not numerous, the harm they do is great. Professional doctors should constantly strive to better their skills and avoid wrong diagnoses and malpractices.

A 19-year-old girl in Hubei Province complained that she could not study. She described agitation, sleeplessness and an inability to pay attention. At a local hospital, she was diagnosed as suffering from a mental disorder linked to puberty. Treatment based on this diagnosis proved fruitless. Her case was reexamined by several experts, who recognized the symptoms of epilepsy. Treatment with herbal medicines alleviated the symptoms, and after five months she was once again living a normal life.

Patient must choose a good doctor for treatment. Meanwhile, doctors must maintain the tradition of stressing professional skills and morality. As ancient people had said, "People are innocent. But they die of mistreatment instead of illness. A doctor should not have learned medical science if he is not proficient in his skills."

## Chapter 17

# Solidarity

Generals and soldiers being of one heart and one mind is one of the most important factors in any victory. To accomplish this point, the most important thing next to strict discipline is to build up the general's prestige among his soldiers. This usually depends heavily on the general's past achievements and current behavior. Those who are farsighted, prudent and victorious find earning their soldiers' respect is not difficult.

Though the relationship between a doctor and a patient differs from that between a general and his soldiers, a wise doctor should take care to be strict with his patients so that better results can be achieved, and medical knowledge spread and popularized.

Sun Simiao believed a doctor should treat his patients as if they were his relatives, going all out to cure the patient without thinking of personal hardship or gain. He said, "A good doctor must be calm and free of any distracting desires. Whenever a person needs treatment, he must not consider whether the person is rich or poor, old or young, a friend or an enemy; and he must not think about his fame or reputation." These are the preconditions for becoming a good doctor.

In addition to treating illnesses, doctors should also concern themselves with a patient's dietary and mental health. Otherwise, the treatment of illnesses will not have good results.

*The Yellow Emperor's Canon of Internal Medicine* says that if a doctor does not dare to be strict with his patients, and if the patients do not follow the doctor's instructions without question, the patient's *qi* and blood will be in disorder, and the patient will not recover.

A doctor cannot stipulate punitive restrictions. But he must take measures to ensure his instructions are followed faithfully.

Educating people to cultivate good living habits and pay attention to exercise and self-cultivation to prevent disease is also one important duty of a doctor. A doctor must not only be a teacher, but must also be an example of the teachings for the people. Only by practicing what he preaches can a doctor convince his patients of the merit of his words.

*The Yellow Emperor's Canon of Internal Medicine* says, "The sage of antiquity did what he required others to do. This is education without words. This is why the people followed him without question."

There are many examples of patients refusing to follow their doctor's instructions and becoming progressively worse. For example, doctors will require their patients to avoid certain foods when taking herbal medicines. One young man was able to cure a mental disorder by taking herbal medicines. He later married and had a son. With time, he began to relax his treatment. His doctor had strictly forbidden him to take any liquor. Once he drank some wine at a wedding. The next day he had a relapse and was readmitted to the hospital. Although he did recover over time, he had destroyed his marriage and lost his son.

Another example is of a teenager suffering from epilepsy. He began a course of herbal medication and began to recover. Although his doctor had warned him not to eat dog meat, he was served some at a relative's home. An hour later he went into a very severe seizure.

Chapter 18

## Observe the Situation

"Assessing the enemy", a term used in Sun Zi's *The Art of War*, refers to the gathering of information about the enemy.

Sun Zi thought that a war was an organic entirety and that the strategic plans and methods of battle of both combatants were inevitably available to the observer, whether directly or indirectly. Victory was a direct function of the commander's discernment.

Emphasizing knowing one's enemy and knowing oneself, he offered 32 methods of assessing the enemy. For example: "If his emissaries sound humble, and yet he steps up his readiness for war, he plans to advance. If their language is belligerent, and they put on aggressive airs, he plans to retreat. If his light chariots move into positions along his flank, his formations are readying to move out. If he has suffered no setbacks, and yet sues for peace, he intends deception. If his troops rapidly close the field, and his chariots remain in the main formation, he anticipates a decisive battle. If some of his troops retreat, while others advance, he is seeking to draw you into a trap. If his soldiers lean upon their lances, they are hungry and tired. If those sent to fetch water drink before filling the storage bladders, his troops are thirsty. If there is an advantage to be gained by pressing the battle, and yet they do not advance, they are weary."

The basic principle of these methods is to see beyond the appearance and get to the truth that lies below the surface.

Historical experiences have proven that those who excel at assessing the enemy are able to seize the day and win the battles.

In 615 BC, the State of Qin attacked the State of Jin. The king of Jin appointed Zhao Dun to fight at Hequ (modern-day Yongji in Shanxi Province). Knowing the Qin troops had traveled on a long expedition and thinking they could not last long, Zhao adopted a strategy of confrontation. He expected to weaken the Qin troops and then to pursue their retreat and defeat them. When faced with overwhelming forces, the Qin leaders decided to withdraw before the battle and regroup. To conceal their intentions, the Qin generals sent an emissary to the Jin headquarters. In a belligerent manner, the emissary boasted of the decisive battle in which the Qin forces would sweep aside the Jin army and march on the capital. One Jin officer recognized the deception and urged his superiors to set a trap behind the Qin forces. Unfortunately, Zhao was not familiar with Sun

Zi's warning — "If his emissaries' language is belligerent, and they put on aggressive airs, he plans to retreat." Zhao did not take his officer's advice. The Qin troops withdrew to safety the night before the battle.

Sun Zi's methods of assessing the enemy greatly influenced generations to follow, not only in the military, but also in the medical sciences.

An ancient doctor once explained, "If a general mechanically applies strategy and tactics without the ability to adjust to changing situations, he will undoubtedly loose the battle. And if a doctor mechanically treats symptoms, never seeking the cause of the illness and learning how to treat disease from this aspect, he will never cure the patient."

Just as Sun Zi saw war as an organic entity, doctors of TCM also consider the human body in a holistic manner. They believe that all parts of the human body are connected by main and collateral channels, through which vital energy circulates and along which the acupuncture points are distributed. They also believe internal diseases are inevitably reflected externally in, for example, the face, tongue and pulse or the spirit.

For example, if a patient has a red face, he is believed to have intense heat in his heart. If his face is pale, he has an infection in his lungs. If his skin is yellow and withered, his spleen is weak. If his face is gray, he is suffering from liver complaints or pains somewhere in the body. People who are excitable or frequently worried have excessive liver-*qi* (liver vigor), and those who are depressed without reason are short of lung-*qi*.

Doctors differentiate between 28 pulse conditions, each reflecting a different disease. For instance, the pulse of a patient suffering from liver disease feels like the string of a musical instrument. That of a patient suffering from a disease of the lung feels like a splinter of wood floating on water. A person suffering from a diseased spleen has a weak pulse.

Doctors of TCM use different diagnostic methods than do their Western counterparts. Without instruments or elaborate tests, Chinese doctors discern information using their own senses. They then analyze this information and make a diagnosis.

Through years of practice, Chinese doctors have summarized four effective diagnosis techniques. These are looking, listening, questioning and feeling the pulse.

By *looking*, the doctor observes the patient's spirit, color, coating on the tongue and excrement. For example, obese people usually suffer from phlegm dampness and therefore should pay attention to preventing apoplexy. Thin people, if they do not suffer from a stomach disease, should have their blood sugar level examined.

Observing a patient's expression, especially his eyes, is quite important to making a diagnosis. Chinese doctors believe the conditions of all internal organs of the human body are reflected in the eyes. For example, if a patient looks terrified, i.e. eyes wide and pupils dilated, he must be suffering from sharp pain. And if a critically-ill patient suddenly becomes fresh and his eyes bright, it is quite possible that he is having a momentary euphoria just before death. But, if a patient's eyes shift, and he frequently swallows, he is most likely lying about his condition.

By *listening* to a patient's description of his own conditions, as well as his breathing, coughing or groaning, the doctor can also learn a great deal. For example, if a patient talks too much or stays too quiet, the doctor would probably gather that he is suffering from liver-*qi* depression, while a Western doctor would diagnose a nervous disorder. If the patient's voice is subdued, he must be very weak.

Meanwhile, smelling and observing a patient's breath, phlegm, nasal mucus and other discharges are also important. For example, if a patient's discharge is glutinous and smells foul, he is suffering from a heat-related syndrome. If the opposite is true, he suffers from a cold-related syndrome.

By *questioning* the patient, the doctor can discover what other illnesses the patient has suffered, when and how they came about and were treated, how he feels now, and changes in his condition after being medicated, as well as his lifestyle.

When a doctor talks of *feeling the pulse*, he is actually referring to a wider spectrum of tests, including checking the pulse, skin, chest and stomach. By touching a patients skin, the doctor can gauge temperature, dryness and smoothness. Probing the stomach aids a doctor in deciding if the cause of the illness is a tumor. The information gathered in this way is pooled and evaluated in order to aid the doctor in making a diagnosis.

The occurrence and development of a disease is a complicated process. In order to gain accurate information about the illness at an early stage, the doctor cannot rely exclusively upon any one of these four methods. Therefore, doctors of TCM stress the comprehensive use of all four.

Any doctor, no matter how extensively trained, who is not skilled at looking, listening, questioning and feeling the pulse, should not be considered qualified. *The Yellow Emperor's Canon of Internal Medicine* says, "Looking at a patient's color and knowing what disease he has, the doctor is a miracle worker; feeling the pulse and knowing what disease he has, the doctor is divine; and questioning his conditions and knowing how he became ill, the doctor is skillful."

Although the four methods have different advantages, they each also have their own shortcomings. Only by combining them can a doctor make use of the advantages and avoid the shortcomings. That's why Chinese doctors stress their comprehensive use.

In the rush to find a speedy diagnosis, it is easy for a doctor to only feel the pulse, ignoring the other three methods. However, this will only result in an inaccurate and lopsided diagnosis. Care should be taken to avoid the expedient when it comes to matters of life and death.

At the same time, the doctor must equally weigh the information gained through each method, distinguishing the changes and nature of the illness so as to allow for an accurate diagnosis. Special attention must be paid to cases where the information from one method contradicts the others.

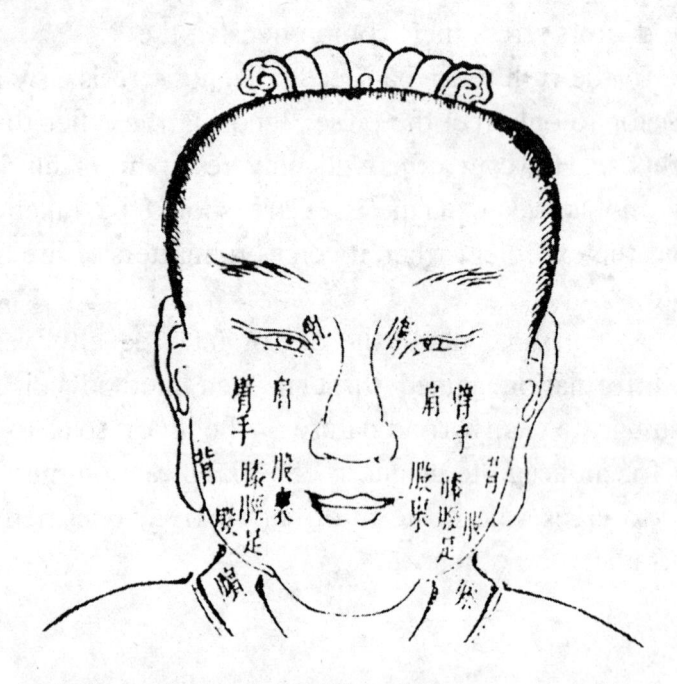

肢节病变见面部　明代医书

Indications of morbid conditions of the limbs on the face (taken from a medical book of the Ming Dynasty).

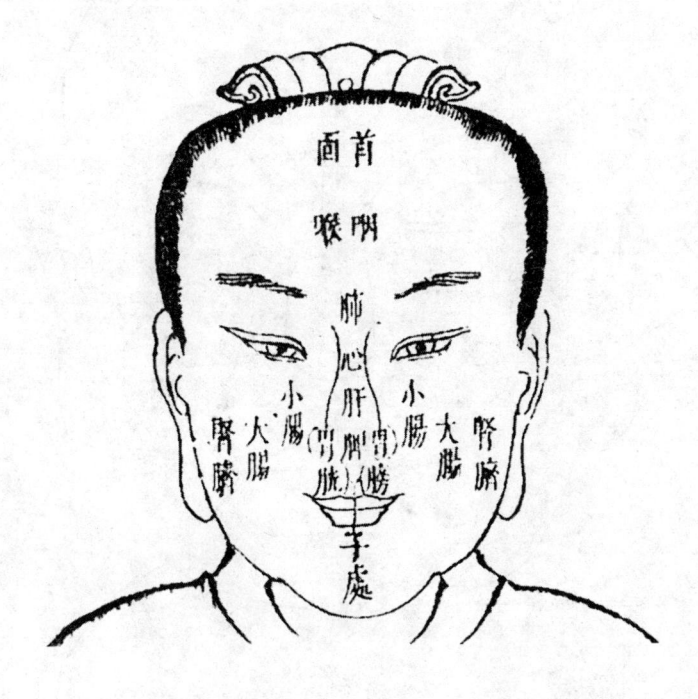

脏腑病变见面部
Indications of morbid conditions of the internal organs on the face.

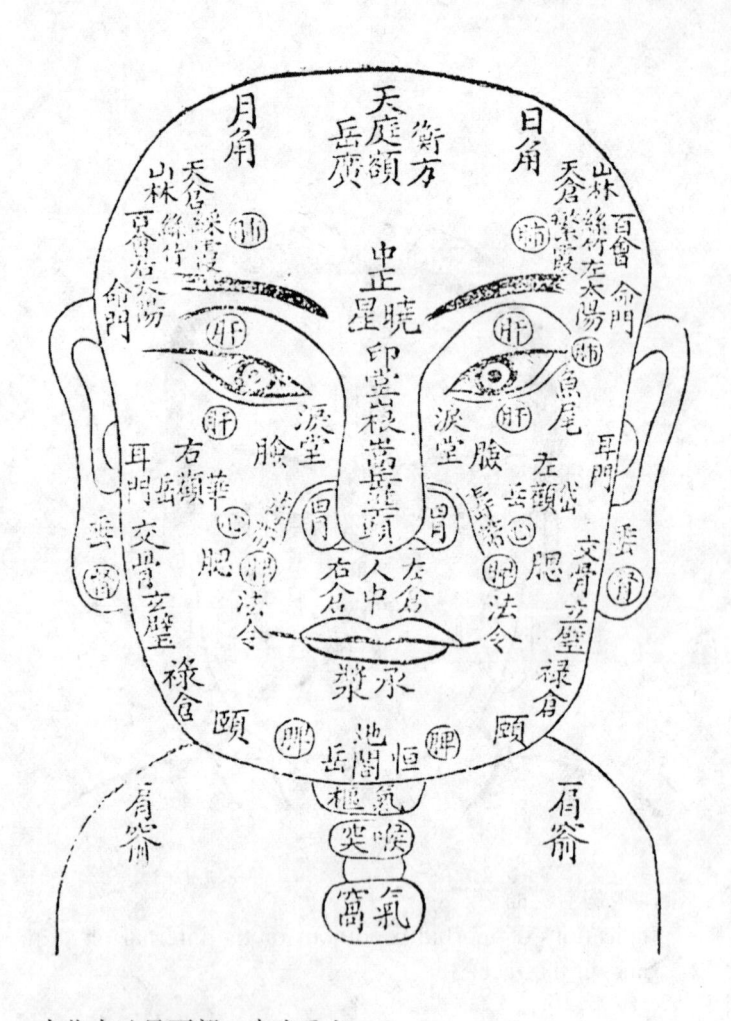

小儿吉凶见面部　清代医书

Indications of physical health of a child on the face (taken from a medical book of the Qing Dynasty).

# Chapter 19

## Analysis Important

The period between 1644 and 1911 was one of renewed interest in Sun Zi's *The Art of War*. Hundreds of books and essays were published on the subject of his writings. One said, "Saving people from war is like saving to a patient from illness. A good doctor writes prescriptions according to a patient's condition. And a good general plans a battle according to the enemy's situation. The 13 chapters of *The Art of War* are actually prescriptions for treating disease."

This commentator believed common points could be found between the conduct of warfare and the practice of medicine.

TCM is different from Western medicine. The latter stresses detailed and concrete research and analysis in such things as viruses and bacteria, while the former stresses holistic and overall research in the factors of illness.

*The Yellow Emperor's Canon of Internal Medicine* says, "Diseases originate from dryness, humidity, cold, heat, wind and rain, or *yin*, *yang*, joy and anger, as well as habits in food and living."

Zhang Zhongjing also said, "There are three reasons for illness. First, main and collateral channels are hurt so that internal organs get sick. Second, the blood vessels and other channels connecting arms, legs and orifices are obstructed. And third, sexual intercourse, household utensils, contact and animals may all bring harm to the human body."

TCM emphasizes diagnosis and treatment based on an overall analysis of the illness and the patient's condition.

This is something like devising strategies for battle. For example, Sun Zi advocated a thorough investigation of the basic conditions of the enemy, including politics, economy and military readiness, so that a general knowledge about the enemy could be gained.

It is the same with TCM. It is not centered on surgery or injections. Instead, it centers on herbal medicines. Suiting the medicine to the illness, traditional Chinese medicine quickly relieves the patient from his sufferings.

Analyzing symptoms by looking, listening, questioning and feeling the pulse, finding the reasons and essence and then making a diagnosis and offering treatment are the mechanisms of TCM.

Diagnosis is the culmination of a period of investigating reasons and symptoms, analyzing their internal connections and finding out what is really wrong. It is like investigating the conditions of the enemy, and then analyzing and judging the information so that proper decisions can be made in battle.

The following example can well illustrate this point.

During the Ming Dynasty (1368-1644) lived a doctor named Lu Lianfu. One evening he returned home tired, with a slight temperature, and aches throughout his body, although he had no headache. He took it for granted that he had become ill due to external factors and took some notopterygium soup to treat the symptoms. However, after drinking the medicine three times, the fever had still not reduced. He then took five or six doses of bupleurum chinense soup. Instead of reducing the fever, his temperature rose. He had to call upon another physician, Yu Tuan.

Yu, seeing the bupleurum chinense soup, felt Lu's pulse and decided Lu was not suffering from a cold syndrome. He warned Lu, "You almost killed yourself with that medicine. You are suffering from internal hurt and weakness. If you drink all this soup, you will die."

But Lu did not agree with him and protested, "I am full of vitality and have never suffered from internal weakness. I am sure I have a cold."

"But your pulse does not feel heavy. And you do not have a headache, and your nose does not feel dry.

Besides, you have taken medicines for a cold for eight days. The fever should have come down if it were really related to a cold," Yu said.

Lu had nothing to say and had to agree with Yu, who then provided Lu with herbal medicines for nourishing internal organs and their vitality and told him to take two doses in the evening.

The next morning, Lu did not feel better. He wanted to take his own medicine. But Yu insisted, "Take two more doses. You can condemn me if there is no result after that."

As Yu expected, Lu's temperature came down after he took the two doses of herbal medicines. Then, Yu, reducing certain items in the prescription, told Lu to take an additional 20 doses so that his vitality could be improved.

Having recovered, Lu said shyly, "I almost killed myself."

In this case, Lu made a diagnosis based only on surface conditions, thus prescribing the wrong medication. Yu, on the contrary, first made clear whether Lu was suffering from an illness caused by internal or external factors and then gave the right medication, which proved effective.

So, the conclusion can be drawn that traditional Chinese medicine shuns simply treating the symptoms and stresses looking beyond appearances to get to the essence of the illness and to diagnose the problem and treat the illness based on an overall analysis of the patient's condition.

# Chapter 20

## Food vs. Medicine

Most medicines have side effects and taste bad, especially herbal medicines. Is there any other way to treat the diseases?

Sun Simiao answered the question with military principles. He said, "A doctor must get to know the causes of a disease and treat it first with food. If this fails, use medicines, because medicines are like fighters, fierce and strong, and should not be used rashly."

He thought that an army winning a battle but taking heavy casualties was like a medication. Although the disease is eliminated, the human organs are harmed to some extent. Thus medication should be seen as a last resort in the treatment of a patient.

He believed a doctor would be thought of as good if he knew how to make people feel relieved as he cured their illnesses with food.

This is the same as the military thought of "trying peaceful means before resorting to force".

The ancient Chinese said, "an army is founded to eliminate violence. It is used when there are no other alternatives. And medicines are used to eliminate illness. They should not be used unless they are absolutely necessary."

Treating diseases with food is like solving political problems through diplomacy. There will be no side effects or complications.

The history of treating illnesses with food can be traced to the 21st century BC. At that time, people found that wine could help stimulate the circulation and relax muscles and joints. In addition to wines, they found that teas, turnips and onions could be used as medications.

In the Song Dynasty (960-1279), a high-ranking court official came down with a serious cough. Local governmental officials sought doctors. But no one wanted to treat the man, fearing his anger if they failed. Finally, an old village doctor, who was also suffered from a serious cough, agreed to examine the court official. On the way, the old man was thirsty and went to a farmhouse to ask for some water. The host gave him a bowl of hot water. Having drunk the water, the old man felt relieved from his cough. He then asked for another bowl of the water and felt better after drinking it. He thought this strange and asked, "What kind water is this?" The host said, "I am sorry. I have no tea. So I provided you with turnip soup."

Full of gratitude, the old doctor asked the host for a bag of dried turnips. In the following couple of days, he cured his cough by eating the dried turnips.

Upon his arrival at the sick official's house, he made his diagnosis and confirmed that the official was suffering the same disease he had just cured himself of. He wrote a prescription and told the official that he must boil the herbs himself. As he boiled the herbs, he secretly added the turnips to the pot. A few days later, the official recovered. With great joy, he rewarded the old doctor with 1,000 *liang* (a *liang* equals 50 grams) of gold.

So, turnips were found to eliminate phlegm, stop a serious cough, guide lung vigor and relieve chest congestion.

*The Yellow Emperor's Canon of Internal Medicine* says, "Proper arrangement of the five tastes (sweet, sour, bitter, pungent and salty) in a diet makes bones set, muscles pliable and tough, and blood and *qi* circulate smoothly." This was regarded as a principle of food treatment by later generations of doctors.

The five tastes are closely related with human internal organs. *The Yellow Emperor's Canon of Internal Medicine* says, "Sour is related to the liver, bitter to the heart, sweet to the spleen, pungent to the lung and salty to the kidney.... People who suffer from liver diseases should not eat hot foods; those who suffer from heart diseases should not eat salty foods; those who suffer from spleen diseases should not eat sour foods; those who suffer from kidney diseases should not eat sweet foods; and those who suffer from lung diseases should not eat bitter foods."

Thus, it can be seen that the five tastes exert great influence upon a person's *qi*, blood and mood.

Proper arrangement of the five tastes has two meanings. One is to avoid a partiality for particular kinds of foods, and the other is to stress different foods at different times.

A partiality toward a particular kind of food usually causes an imbalance of the internal organs, thus inducing disease. For example, more fat than necessary will induce arteriosclerosis or coronary heart disease. And a lack of plant fiber will influence the metabolism of the stomach. That's why *The Yellow Emperor's Canon of Internal Medicine* tells people, "Grains maintain human health, while fruits help; meat is good, and vegetables are a supplement. Eating a combination of these cultivates health and vital energy."

The five tastes must be well arranged all the year round because changes of seasons have direct and indirect influences upon the internal organs. Eating more or avoiding certain foods in different seasons helps balance the internal organs and smooth *qi* and blood. For example, hot foods stimulate the appetite and body temperature, making people feel warm. This is beneficial during winter. In summer, a person is better served by eating cold dishes.

Traditional Chinese medical beliefs hold that medicines and food are of the same origin. That is to say that there is no strict demarcation between food and medicinal treatments. Stressing one does not mean eliminating the other. On the contrary, the two methods should complement each other. When one is slightly ill, he should emphasize changes in the diet. But when one is seriously ill, he should emphasize medicinal care.

## Chapter 21

## The Significance of Tactics

"Treating a disease is like dealing with enemies." This is a famous saying of TCM. The ancient Chinese believed that "if you know where the enemies are and attack them with crack troops, you can win the battle without staining your sword."

"Winning a battle without staining your sword" is the highest pursuit of the ancient Chinese militarists. A famous point of view of Sun Zi was that "to fight 100 battles and win each and every one is not the wisest thing to do. To break the enemy's resistance without fighting is."

These words are also proper in traditional Chinese medicine.

The mainstay of TCM is herbal medicine. By administering herbal medicines according to indications, illnesses can be eliminated with a few doses.

For example, Hua Tuo, 141-203, once received two patients, Ni Xun and Li Yan, who were both suffering from headaches and fever. Hua Tuo made careful diagnosis, wrote two prescriptions and said, "Ni should discharge, but Li should promote sweating."

The two patients thought it strange and asked, "We are suffering from the same disease. Why do you treat us in different ways?"

Hua Tuo smiled and told them, "Ni is suffering from external excess syndrome, and Li is suffering from internal excess syndrome. That's why I am treating you differently."

Ni and Li took the herbal medicine prescribed by Hua Tuo. By the next day, they had both recovered.

It seems ridiculous that Hua Tuo's diagnosis could be explained in a few words. Of course, his diagnosis was the result of a long period of examination and observation. He had to know where the *enemy* was and then to find the *crack troops* to *attack* the enemy.

Over centuries of practice, generations of doctors of TCM have defined a variety of *crack troops*. But prescribing the correct remedy is not always easy. Experience and skill are needed.

Of course, there are routines to follow. Ancient doctors summarized eight strategies: sweating, vomiting, discharging, warming up, harmonizing, nourishing, clearing and removing.

External pathogens first harm human skin and then move into internal organs. When the pathogens are in the skin, they may be eliminated through sweating. But this method does not suit a patient who has lost a great deal of blood or one who vomits or has diarrhea.

Vomiting refers to removing phlegm, retained food, or poisons from the stomach. When people vomit, they also sweat. This also helps eliminate external pathogens. This method is useful for patients suffering from serious retention of dominant evil. The elderly and the weak, as well as pregnant women, should avoid this method.

Discharging is relaxing the bowels, eliminating excess heat or releasing blood stasis. The elderly and the pregnant should avoid this method.

To harmonize is to boost vital energy and eliminate pathogens by means of adjustment. This method is mainly used when the illness is between internal organs and muscles and skin and cannot be sweated out or discharged.

Warming is meant to eliminate cold, smooth the flow of blood and boost spleen and kidney actions.

Clearing refers to removing excess heat and cooling the blood.

To remove is to eliminate indigestion, extravagances, phlegm and humidity. This method is used when the illness is among internal organs, vessels or muscles. Obstinate but mild, these types of illnesses, especially when caused by accumulation of *qi* and blood, cannot be eliminated immediately. They must be cleared gradually.

Nourishing is the boosting of the energy and blood

and the cultivating of vitality. Any weakness caused by deficiency of *qi*, blood, *yin* or *yang* can be treated by this method.

Ancient doctors thought that these eight methods constituted the eight trunks of treatment of traditional Chinese medicine, with thousands of branches joining each trunk. A doctor must be able to find the right combinations of branches by observing the patient's condition.

They believed that because the condition of the patient may change, proper treatment can only be found if the doctor has achieved a mastery through comprehensive study.

# Chapter 22

## Formulas and Formations

Writing a prescription is like arranging a battle formation.

A battle formation arranges soldiers so that they can play their best role on the battlefield. After the battle begins, the formation must be adjusted according to the changing situation.

For example, a single cavalryman is no match for an infantryman on a level battle ground. But when arranged in formations, the cavalrymen are able to defeat enemies of surprisingly higher numbers.

Aware of the spirit of this idea, the medical scientist Zhang Jingyue, 1562-1639, divided 1,516 ancient prescriptions into *The Eight Formations of Ancient Prescriptions* and another 186 contemporary prescriptions into *The Eight Formations of New Prescriptions*.

The eight formations were nourishing, harmonizing, attacking, discharging, cooling, warming, consolidating and analyzing.

Nourishing prescriptions are mainly used for languid and weak people. When weak in $qi$, use ginseng and the root of membranous milk vetch; when weak in vigor, use Rehmannia glutinosa and Lycium chinense; when weak in *yang*, use cassia bark, monkshood and dried ginger; and when weak in *yin*, use Ophiopogon japonicus, Chinese herbaceous peony and dried rhizome of rebmannia.

The purpose of harmonizing is to adjust human internal organs and *yin* and *yang*. When ill and weak, a patient should be nourished and harmonized; when ill and stagnant, he should be discharged and harmonized; when ill and cold, warmed and harmonized; and when ill and hot, cooled and harmonized. Combinations of medicines that have opposite functions should be avoided.

Attacking prescriptions are used for acute, excess diseases, which are serious like strong enemies. For example, attacking concentrated *qi* helps smooth vigor energy, and attacking extravasated blood helps smooth blood circulation. With a patient ill in *yin*, medicines attacking *yang* should not be used; and with one ill internally, medicines attacking the surface should not be used. Although this method eliminates illness, it also brings harm to other parts of the body. So, it should not be used unless necessary.

Discharging is used to treat illness in muscles and skin. When using discharging medicines, a doctor must know their properties, because some medicines are strong, some are moderate, and some are weak.

Cooling medicines are used to treat heat diseases. In the same way, a doctor must know the properties of these medicines because some medicines, such as Scutellaria baicalensis and Dendrobium nobile, clear heat in the upper part of the body, while others, such as Cape jasmine and rough gentian, clear heat in the lower part of the body.

Warming medicines are just the opposite of cooling medicines. For example, dried ginger warms internal organs and helps eliminate illness in muscles and the skin; Chinese cassia invigorates the circulation of blood; and nutmeg warms the stomach.

Consolidating medicines are used with patients who suffer from such diseases as diarrhea and chronic asthma. For example, the lungs should be consolidated when a patient suffers from chronic coughing; the bladder should be consolidated when a patient cannot control urination; and when a patient has loose bowels because of cold, his bowels should be consolidated with warming medicines. In short, to treat diseases in the upper part and surface of the body, *qi*, which is centered in the lungs, should be consolidated; and to treat diseases in the lower part and internal organs, vital essence, which is centered in the kidneys, should be consolidated.

Analyzing methods entails prescribing treatments according to each patient's conditions. For example, when there is a carbuncle, swelling must be eliminated. And when a fracture has occurred, it must be rejoined.

The eight formations of prescriptions stress suiting the medicines to the illness. At the same time, the medications must be coordinated with each other and also flexible.

Xu Dachun captured the essence of the eight formations when he said, "moderate medications, instead of strong ones, should be used when a patient is weak, and the condition is chronic. But when a patient is strong, strong medications should be used at the beginning of the treatment, and moderate medications should be used to assist the result. This is also true in the survival of a nation. When it is weak, it should not drain its manpower and material resources. But when it is strong, it can fight."

# Chapter 23

## Flexibility

Xu Chunfu, a noted physician in the Ming Dynasty (1368-1644), said in his work, *Ancient and Modern Medicine*, that curing a patient is like fighting a battle — the best strategy to counter an offensive is with a force commensurate with the strength of the enemy. In medicine, he said, the best remedy is to apply the right medicine for the right disease. This analogy is indeed pertinent because flexibility is important in both medicine and warfare.

Sun Zi, in his *Art of War*, said, "Generally, in battle, use normal, regular tactics to engage the enemy, and use unusual and unexpected ones to score a victory." By this, he meant that in battle, the normal tactic is to use the main force to attack the enemy head-on and use a support force to attack the flanks.

Sun Zi further said, "When you outnumber the enemy ten to one, surround him; when five to one, attack him; when two to one, divide him; and if equally matched, stand up to him. If you are fewer than the enemy in number, retreat. If you are no match for him, try to elude him."

Tactically, if one outnumbers the enemy five to one, then one should attack the enemy head-on, with the main thrust, which should be three times the number of the enemy, while the support force, which should be two times the number of the enemy, attacks the flanks.

Xu applied this tactic in medicine and proposed that when performing acupuncture and prescribing medication, the doctor should take into careful consideration the specific conditions surrounding the patient.

Traditionally, Chinese doctors paid meticulous attention to the compatibility of medicines when giving a prescription. This theory was first developed in *Shen Nong's Materia Medica*. The author classified the 365 herbal medicines listed in the book into four categories and named them *jun* (monarch), *chen* (minister), *zuo* (assistant) and *shi* (envoy) to indicate their functions.

*Jun* refers to the medicines that are supposed to play the leading role; *chen* refers to those used to reinforce the main medicine; *zuo* refers to the auxiliary medicines to alleviate the toxic effects of the main medicines or corrigents; and *shi* refers to the medicines that guide the main medicine to the site of the illness.

The book explains the main functions, properties, tastes and origins of every herb and provides a clear reference for doctors as to the compatibility of medicines.

Apparently, *jun* is comparable to the "regular

force" that launches frontal attacks, and the other three categories are similar to the "irregular force" responsible for auxiliary attacks. There are many cases in TCM that bear witness to the fit between therapy and warfare.

A typical example is the concept of "routine treatment" and "non-routine treatment". The former is treatment with recipes or drugs opposite to the nature of the disease. For example, drugs of a hot nature can be used for cold syndromes. These cold syndromes can be divided into two types: external cold, such as common cold of pathogenic wind-cold type; and internal cold, a degradation of the function of an internal organ or the whole body, marked by chills, diarrhea, dropsy or impotence. Drugs prescribed to cure cold syndromes must be of a hot or warm nature, most having a hot taste. Perillaseed leaves and scallion stalk, for example, are of a hot nature and capable of curing the common cold. Dried ginger can treat a stomachache caused by cold factors. Lizhongwan, a drug formulated to regulate the functions of the stomach and spleen, is made primarily of dried ginger.

On the other hand, drugs of a cold nature can be used to dispel heat and cure heat syndromes. Like cold syndromes, heat syndromes can also be divided into two types: superficial and interior. Superficial heat is most often seen in the common cold, marked by thirst and a sore throat. Interior heat refers to sthenic heat of the stomach and the lung or retention of the heat pathogen in the liver and gallbladder, marked by fever without chill, a bitter taste in the mouth, restlessness, oliguria with reddish urine, swelling of the gum and constipation. Drugs used to cure heat syndrome are of a cold nature with bitter taste. These include mulberry leaves, chrysanthemum, and goldthread.

In addition, tonics are invigorative of weaknesses, such as panic, timidity, insomnia and amnesia. These can be treated by Tianwang Buxin Wan (Cardiotonic Pill). Paleness, weakness and thin excretion are symptoms of a weak spleen and can be treated with Shenling Baizhu Wan (Powder of Ginseng, Poria and Bighead Atractylodes). Back pains, cold limbs and impotence, on the other hand, are symptoms of a weak kidney that can be treated with kidney-invigorating pills.

Likewise, holagogue can lead to diarrhea, a method known as "treating sthenia syndrome with purgation". Sthenia syndrome refers to the "extra" elements in the human body, such as germs and toxic elements that lead to disease. Food that cannot be digested and discharged also constitutes "extra" elements. These include "dry urine" and urinary retention.

In addition, *qi* and blood, which are normally useful to the human body, can also become toxic, such as blood stasis, if some internal disorder arises.

All these extra elements are called sthenia, and should be discharged.

Other examples of routine treatment include the use of diminutive drugs to treat *qi* dysfunctions; the use of phlegm-eliminating drugs to treat phlegm syndrome; the use of digestive drugs to treat retention of food in the stomach and blood stasis; and the use of parasite killers to treat parasitism.

There are, of course, variations to every disease, and sometimes the symptoms can be deceptive. In this case, the "non-routine treatment" — treatment applied when false signs of a disease appear and the routine treatment is not suitable — should be applied.

For example, in some cases involving fever due to exogenous pathogenic factors, the internal heat reaches a high degree, although externally, the limbs are cold. This is because the cold symptoms are false, and the heat inside is the real problem. Therefore, medicines of a cold nature must be applied.

On the contrary, some cold syndromes may have symptoms such as rosy cheeks, feverish body temperature, thirst and restlessness. These symptoms, however, are misleading because the internal cold is the real illness. Therefore, drugs of a warm or hot nature must be applied.

Sometimes, insufficiency of the spleen, which theoretically should result in diarrhea, appears instead as constipation. In treatment, spleen-invigorating medicines should be applied. This is also a typical non-routine treatment.

In addition, diarrhea caused by retention of food should be treated not with the anti-diarrhetics but instead, with purgatives.

The above therapies are contrary to the routine treatment in that the medicines are seemingly incompatible with the nature of the disease. Essentially, however, they are just variations of the "routine treatment". They testify to the ancient adage that "medication is like maneuvering troops."

# Chapter 24

## Discretion

A man by the name of Zhang had a lump in his abdomen, which resulted in retention of a mass in his chest and difficulty in digestion. Desperate, he consulted a doctor. After diagnosis, the doctor told him that purgation had to be used to remove the stagnation of his internal organs. He then gave the patient a prescription.

After returning home, Zhang wondered whether a little overdose would accelerate the treatment. He then proceeded to drink all the concoction. To his amazement, the symptoms disappeared the next day, and everything returned to normal — his digestion, breathing and feelings. Several days later, however, the old discomfort returned. Zhang drank a dose greater than the last time, and the effects were even better.

In the following month, Zhang had five bouts of the lump, and each time it was subdued only by a greater dosage. In the meantime, Zhang was exhausted, having difficulty breathing. He sweated profusely, with involuntary trembling. Although he performed no strenuous labor, Zhang felt weak and listless all day long. He wondered why the illness persisted.

Later, he heard of a great doctor in a neighboring country and went to seek his help. The doctor told him the problem did not lie with the illness, but with the way he took his medication. "Since your illness is caused by the dysfunction of the heart and the stagnation of *qi* and blood — a dangerous symptom indeed — it is totally inappropriate to use a large dosage for treatment, although that would produce an expedient effect," said the doctor.

"This is because of the *qi* of harmony circulating in the body cannot withstand the onslaught of the medication," the doctor explained. "That's why everytime you use a greater dosage than needed the illness seems to go. The truth is that your *qi* of harmony is also hurt along with the exogenous factor. That's why you feel so weak and listless. To treat your lump in the abdomen while preserving the *qi* of harmony, you'd better go back and lie down for three months before coming back to get your prescription."

Three months later, Zhang went back to the doctor, who observed that the primordial *qi* (source of life) was coming back, and treatment could commence. He told the patient to take the medicine according to the prescription. "If you do this, then after three months your illness will start to alleviate; after another three months your body will start to recover; and after a year you will fully recover," he said.

Zhang followed the order of the doctor, and everything turned out just as predicted.

After recuperation, Zhang visited the doctor and asked him to explain. The doctor, instead of explaining the medical workings of the question, related the story of the collapse of the State of Qin due to autocratic rule. He concluded that expediency would do no good in the end.

This story was recorded in *Admonitions to Doctors*, compiled by Zhang Lei, a doctor in the 10th century. A major principle embodied in the story is the exercise of discretion when taking medication.

As *The Analects of Confucius* recorded, Confucius was always cautious about three things: offering sacrifices to gods and ancestors; war; and disease. This cautious attitude makes good sense. In fact, ancient Chinese strategists without exception warned against the grave consequences of waging war.

Needless to say, the same discretion was also found in TCM. As *Assorted Prescriptions of Materia Medica* points out, "Prescribing medication is like meting out punishment — a mistake will make it a matter of life and death."

Usually, the patient and their family panic over disease and wish to recover as soon as possible. Most medicines, however, produce side effects and must be taken with caution.

The dosage is not determined at random, but rather is based on three factors. The first is the property of the medicine. As a rule, hazardous or toxic medicines should be taken in small doses, and any increase in dosage should be made in accordance with the need arising from the course of treatment. Once the symptoms are receding, the dosage should be gradually reduced or stopped immediately so as to prevent intoxication or any side effects. For ordinary medicines, those with greater mass, such as minerals and shells, should be taken in large doses; others with light mass such as flowers and leaves, as well as aromatic herbs, should be taken in small doses. In addition, those with a strong and greasy taste should also be taken in large doses.

The second factor for deciding dosage is the compatibility of the medicine with the nature of the disease. Generally speaking, for the same type of medicine, a simple recipe should use a greater amount than a compound prescription, and a decoction should use a greater amount than pills and powders. Furthermore, in compound prescriptions, the principal medicine should have a dosage greater than the adjuvant drug.

The third factor is the personal conditions of the patient, such as the seriousness of the disease, physique and age. For serious, acute and persistent illnesses, the dosage is normally greater than normal; on the other hand, mild and chronic diseases require small dosages. Physically strong patients can use greater dosages than the physically weak, women and children. Also, a fresh disease dictates greater dosages than persistent ones. The following table sets out the different dosages for different ages.

| Age | Dosage |
|---|---|
|  | (proportion to adult's dosage) |
| Newborn-one month old | 1/18-1/14 |
| One-six months old | 1/14-1/7 |
| Six months-one year old | 1/7-1/5 |
| One-two years old | 1/5-1/4 |
| Two-four years old | 1/4-1/3 |
| Four-six years old | 1/3-2/5 |
| Six-nine years old | 2/5-1/2 |
| Nine-14 years old | 1/2-2/3 |
| 14-18 years old | 2/3-full |
| 18-60 years old | Full-3/4 |
| Over 60 years old | 3/4 |

The way to take medicine is also grounds for discretion. Generally, tonics and invigoratives should be taken before meals, while medicines that stimulate the stomach and intestines should be taken after meals. In addition, parasiticides and anti-diarrhetics should be taken on an empty stomach.

Other examples: tranquilizers should be taken before going to bed; medicines for acute diseases should be taken immediately regardless of time; pills, powders, soft extracts and wines for chronic diseases should be applied regularly (such as three times a day or in the morning and evening) so as to let the medicines sink in.

To maintain the best effects, medicines that are prescribed to be taken before or after meals should be taken with a time gap of one to two hours. Usually, one dosage of herbal medicines can be taken two to three times. Those with mild symptoms can take it once in the morning and once in the evening, while those with acute symptoms can take it once every four hours.

When applying medicines administered to induce diaphoresis and remove constipation, the compatibility of the medicines with the physique of the patient must be taken into consideration. Once diaphoresis is induced and constipation removed, medication can be stopped regardless of whether the dosage is finished or not. This is necessary to prevent damage to the vital $qi$.

Li Gao, 1180-1251, a noted physician in the Jin Dynasty (1115-1234), observed, "Disorder in medication is like anarchy in the armed forces — a dauntless general will only make matters worse."

As it is, the reasonable use of the armed forces serves the interest of the state and the people, while excessive use will only result in self-destruction. The same principle applies to medication. Any medicine is prone to create deviation of either the *yin* or *yang* elements. Correct use of medicines can remedy the imbalance of *yin* and *yang* forces within the body, whereas excessive uses inevitably hurt the vital *qi*.

What, then, is the yardstick for measuring the right dosage? *The Yellow Emperor's Canon of Internal Medicine* dictates, "When applying a drug with the greatest toxic effects, medication should terminate once 60 percent of the symptoms are gone; when applying a drug with ordinary toxic effects, medication should terminate once 70 percent of the symptoms are gone; when applying a drug with minimal toxic effects, medication should terminate once 80 percent of the symptoms are gone; and when applying a drug with no toxic effect, medication should terminate once 90 percent of the symptoms are gone. The remaining ailment can be cured by food (cereals, meat, fruit and vegetables) and exercise."

Even tonics, though able to invigorate the vital *qi* of the body, can cause an imbalance of the *yin* and *yang* elements in the body if used for too long. As *Plain Questions*, a chapter of *The Yellow Emperor's Canon of Internal Medicine*, warns, "It is a natural law for tonics to strengthen the *qi*, but too much strengthening is cause for dying young."

# Chapter 25

## The Heart of the Matter

"Heart" was a very important concept in TCM, and since ancient times there has been a distinction between the physical heart and the mental heart. The physical heart regulates the five internal organs (heart, liver, spleen, lung and kidney), six hollow organs (gallbladder, stomach, large intestine, small intestine, bladder and triple energizer) and the blood vessels all over the body. The mental heart, on the other hand, controls the mind, spirit and thinking. This theory was well expounded in a medical textbook produced in the Ming Dynasty. The book describes the physical heart as "shaped like an unblossomed lotus located below the lungs and above the liver" and the mental heart as "controlling everything in the universe". Apparently, the ancient Chinese valued the mental heart more than the physical one.

A military classic written in the 8th century BC contends that a preemptive strike should be directed at crushing the morale of the enemy. Clearly, the ancient Chinese realized that the heart controls the spirit and mind, and that the spirit and mind have a direct bearing on war. During the Spring and Autumn Period (770-476 BC), Sun Zi, the master strategist, took on that idea and advised that "an entire army can be demoralized, and its general deprived of his presence of mind". He further proposed, "The best policy in war is to thwart the enemy's strategy; the second best is to disrupt his alliances through diplomatic means; the third best is to attack his army in the field; the worst policy of all is to attack walled cities."

Sun Zi continued, "War is a game of deception. Therefore, feign incapability when in fact you are capable; feign inactivity when ready to strike; appear to be far away when actually nearby, and vice versa;" and "Seize whatever the enemy prizes most, and he will do what you wish him to do."

These stratagems have for centuries benefited commanders. A typical example is the strategy employed by Ma Su, a general in the Three Kingdoms Period (220-280).

On an expedition to the south, Zhuge Liang, the prime minister of the Kingdom of Shu, consulted Ma on military strategy. Ma advised that the destination of the army was inhabited by minority ethnic groups whose people were known for bravery. Geographically, the region was also too perilous and remote to conquer. Even if it could be subdued temporarily, it would soon become another source of instability, Ma argued. Furthermore, the archrivals of the Shu rulers were the Kingdom of Wei in the north and the Kingdom of Wu in the east. To avoid being attacked front and rear, it was essential for the Shu to pacify the ethnic minorities in the south. The best strategy, he suggested, was to conquer the land by winning over the hearts of the inhabitants of the land, and the worst one was to attack the enemy's walled city.

The prime minister agreed and adopted Ma's suggestion. As a result, he employed both a "big stick and a carrot" and conquered the area.

Xu Dachun, a noted physician in the Qing Dynasty, said in his *Treatise on the Origin and Development of Medicine* that Sun Zi's *The Art of War* served as a complete guide to medicine. Whereas Sun Zi suggested the best strategy for winning a battle is to demoralize the enemy, doctors agree that the best treatment is to restore the normal mentality to the patient.

*The Yellow Emperor's Canon of Internal Medicine* points out, "It is impossible to cure a disease without taking into consideration mental and psychological factors." Another prominent physician in the Ming Dynasty, Li Zhongzi, emphasized in his *Required Readings for Medical Professionals*, that diseases caused by bad moods or emotions are beyond the reach of even the best medicines. Therefore, it is necessary to treat the patient's heart before the body.

Psychotherapy in TCM refers to a therapy without using tangile instruments such as acupuncture, medication or surgery. The therapeutic instruments used include language and behaviors to induce, instruct, console, regulate, or even stimulate the patient in order to help patients better understand their illness, dispel their anxieties and build up their confidence and capabilities in overcoming the disease. *History of the Three Kingdoms*, a historical account of three rival kingdoms between 220-280, recorded a case whereby Hua Tuo, the best-known ancient Chinese physician, treated a patient with psychotherapy.

A local official had been sick for a long time. Hua Tuo, after performing a diagnosis, concluded that the person could purge his long-standing illness if he flew into a rage. Therefore, he took the patient's money, but gave no prescription. Then he left without saying good-bye. He even left a letter humiliating the official. Enraged, the official sent for the doctor, attempting to murder him to vent his outrage. The son of the official, however, aware of the good intent of the doctor, secretly stopped the agents and told them to report to the official that the doctor had escaped. The official, upon learning the news, burned with fury and choked with blood stirring inside. Finally, he vomited a pool of black blood. To his surprise, his illness was gone.

Modern medical science reveals that the human

immune system is regulated by the nervous-endocrine system and is subject to psychological factors such as cognition and emotions. Unhealthy emotions may seriously inhibit the immune functions of the human body, which will in turn decrease the cancer-fighting capabilities of the body and create internal chaos. Psychotherapy is especially recommended for cancer patients.

A case in hand is Qin Yi, a noted film actress in China, who experienced four major illnesses and underwent seven operations. In 1962, she had an operation to remove a thyroma. Four years later she had intestinal cancer, a morbid disease. Undaunted, Qin kept a high spirit and optimism and persisted in doing health-care exercises. More than three decades have elapsed, and Qin is still alive and healthy. Her story has earned her fame as the "cancer-fighting star".

Medical experience has proven that psychotherapy is the best possible option similar to the strategy "subduing an enemy without fighting a battle."

Doctors use a variety of psychotherapeutic methods, including counseling, suggestion, catharsis and *qigong*.

中国古代医生为患者动手术情景　古吴营平画
An ancient Chinese doctor operates on a patient.

# Chapter 26

## Taming Emotions

In ancient Chinese military strategies, a most important prerequisite for being a commander was the ability to exercise self-restraint, especially in major strategic decision-making. Sun Zi advised, " The sovereign should not start a war simply out of anger; the commander or general should not fight a battle simply because he is resentful. Take action only if it is to your advantage. Otherwise, do not. An enraged man may regain his composure, and a resentful person his happiness, but a state that has perished cannot be restored, nor can the dead be brought back to life. Therefore, the enlightened sovereign approaches the question of war with utmost caution, and the good commander warns himself against rash action."

Clearly, personal emotions should never interfere with decision-making involving the welfare and survival of the country and the people. Therefore, it is of paramount importance for the rulers and commanders to remain composed and cool-headed.

Noxious moods and emotions not only disrupt decision-making, but also harm health. Therefore, self-composure is important to health preservation.

*The Yellow Emperor's Canon of Internal Medicine* dictates, "All diseases stem from *qi*. Anger results in the *qi* moving upward; happiness in the *qi* moving slowly; sorrow in the *qi* dissipating; terror in the *qi* moving downward; fright in the *qi* disturbed, and anxiety in the *qi* obstructed." TCM theory maintains that these seven emotions — happiness, anger, melancholy, anxiety, sorrow, fright and terror — can both cause diseases, if an imbalance prevails, and cure diseases, if used in combinations. This therapy uses stimuli and mental regulation to contain or dispel noxious emotions and eventually control and cure the disease. The following two cases illustrate the point.

According to *Classified Medical Records of Famous Physicians*, a book compiled in 1552 by Jiang Guan, a noted physician in the Ming Dynasty, once a prisoner, tied with robes, was escorted by a guard to the county seat. On the way the prisoner drowned himself. After learning of his death, the prisoner's family members filed a suit against the guard for extorting and persecuting the prisoner to death. The court ruled, however, that the guard was innocent, but he had to pay some damages. The guard became haunted with the loss of money and acted like an idiot. Finally, a doctor named Wang Shishan came to his assistance. The doctor concluded that the patient fell sick over loss of money and had to be cured by an overwhelming mood of happiness — the so-called "happiness overcoming sorrow" formula. He told the family members to make several fake silver dollars out of tin and put them beside the patient's pillow while he was asleep. When the patient woke up and, to his surprise, found so many silver dollars beside him, he was overcome with joy and held the "money" tightly to his bosom. His illness was greatly relieved, and not long thereafter, he returned to normal.

*Confucians' Duties to Their Parents*, another book compiled by Zhang Zihe in the Jin Dynasty (1115-1234), recorded a case of a woman, who, while traveling by herself, stayed the night in an inn. It happened that a gang of highwaymen robbed the inn and killed several tenants. The woman was so frightened that she fell from her bed to the floor. Ever since that encounter, she would faint whenever she heard some extraordinary noise. Her family members had to tiptoe at night in order not to frighten her. A year went by, and the woman remained paranoid. Many doctors diagnosed her as having mental problems and prescribed ginseng, pearl and tranquilizers for her. None of those worked, however.

One day, a doctor named Zhang Dairen came by and diagnosed that the woman's gallbladder — an organ that TCM believes affects a person's ability to make decisions and control emotions — had been hurt in her fright, and therapy should aim at reinforcement. He therefore ordered two maids to tie the woman's hands onto a high chair and put a small tea table in front of her. He told the patient to stare at the tea table, which he then struck with a loud bang that made her almost faint with fright. When she regained her composure, the doctor asked her, "What were you afraid of when I banged at the tea table?" Thereafter, he repeated the trick many times. Each time, the patient's fright was mitigated. Then, the doctor knocked at the door with a stick and had somebody lift up the curtain and make faces to the patient. Finally, the patient was no longer surprised at the tricks. Thereafter, she was no longer frightened even during a thunderstorm.

Asked what kind of therapy this was, the doctor said that *The Yellow Emperor's Canon of Internal Medicine* taught that the frightened should be calmed down. He explained that he was trying to dispel the fear in the patient by repeating what she was fearing.

In addition to these cases, TCM also uses other psychotherapeutic methods to treat emotional disorders.

# Chapter 27

## Diversion

The history of war in both China and abroad suggests that a key factor to victory is to take the initiative in one's hand and outmaneuver the enemy. Sun Zi advised that the "skilled in war move the enemy rather than be moved by him." He then proposed a number of tactics, including one of changing the direction of the enemy's attack. "When we wish to give battle, the enemy cannot but leave his position to engage us even though he is safe behind high walls and deep moats, because we attack a position he must rescue. When we wish to avoid battle, we may simply draw a line on the ground by way of defense, and the enemy cannot engage us because we have diverted him to a different target," he said.

Sun Bin, a descendant of Sun Zi, employed the tactic and masterminded what came to be known as the "battle to rescue the State of Zhao by besieging the State of Wei". In 354, Wei besieged and attacked Handan (in modern-day Hebei Province), capital of Zhao. The next year, the king of Zhao sent an envoy to the neighboring State of Qi for help. The king of Qi dispatched a contingent headed by Tian Ji and Sun Bin to rescue Zhao. Sun, realizing that the State of Wei itself must be loosely defended since most of its elite troops were deployed around the capital of Zhao, decided to attack the capital of Wei instead. Pang Juan, the commander of the Wei army, rushed back to defend his country's capital. Sun then ambushed the Wei army en route, defeating it and capturing Pang.

In therapy, a similar treatment is used to divert the attention or anxiety of the patient from his or her illness to something or somebody else, thereby changing the patient's mistaken beliefs or unhealthy habit. This is well illustrated in the following case.

Once upon a time there were two doctors by the names of Zhang and Wang. Dr. Zhang, confident of his medical expertise, looked down on Dr. Wang. However, it turned out that visitors flocked to Dr. Wang, while Dr. Zhang was barely noticed. Dr. Zhang, seeing the bustling business of Dr. Wang, got so jealous that he could not eat or sleep. Finally, he fell sick. When fellow physicians came to visit him, they invariably asked him how his business was, which only aggravated his sickness. None of the precious medicines he was taking worked, and his conditions deteriorated. Out of resignation, his son suggested that Dr. Wang be invited over to treat him. The enraged doctor flatly rejected the suggestion, saying it didn't make any sense since his illness was caused by that doctor.

However, since his sickness became worse day by day, Dr. Zhang eventually agreed to invite Dr. Wang over. Dr. Wang knew exactly what caused Dr. Zhang's sickness. He pretended to "take the pulse" for his patient by placing his fingers on the side of the bed and announced that the patient's illness had been caused by an irregularity of menstruation. He then gave a prescription to Dr. Zhang. Seeing the prescription, Dr. Zhang shook his head in disbelief. Dr. Wang then suggested applying an adhesive plaster and proceeded to heat it over a lamp. Dr. Zhang exposed his back in preparation for the application. Dr. Wang, while telling the patient to keep himself warm, applied the plaster on the wall by the patient's bed. He promised that the patient would feel better that night, get off bed the following day, go outdoors two days later, and fully recover in seven days.

After Dr. Wang left, Dr. Zhang chuckled. "This Dr. Wang," he said, "is just a quack. Why? Nobody would take the pulse by touching the side of the bed; menstrual irregularities occur to women only, and I'm an old man; even if I had really had irregular menstruation, he should have given me medicines such as Chinese angelica root, white drug-powder, safflower and red sage root, but instead he prescribed medicines that are normally used to accelerate milk production. Also, only an idiot would apply the adhesive plaster onto a wall."

After the experience, he related the story to everyone he met, laughing and gesticulating all the time. Seven days later, as Dr. Wang predicted, Dr. Zhang felt the stagnant *qi* inside his body was gone, his appetite returned, and he could move about freely. Only much later, however, did he realize the purpose of Dr. Wang's therapy and start to admire Dr. Wang from the bottom of his heart.

A contemporary case reported in a 1984 issue of the magazine, *Longevity*, also testifies to the validity of the diversion therapy. Sun Jingxiu, a great children's storyteller, suffered from internal injury from an arm wrestling incident in his childhood, and as a result, he threw up blood at three o'clock every morning. His mother could not afford to get him a doctor or buy invigoratives for him. Seeing him getting weaker day by day, his mother burned with worry for him. One day, seeing the clock would strike three, his mother was suddenly inspired to turn the clock one hour faster. A moment later, Sun woke up, only to find that the clock had struck four. Curious, he asked his mother why he had not vomited blood yet. His mother answered, "My son, your illness is gone. You see, it's four o'clock already, and you're fine." From then on, Sun, now rid of mental torment, became better and better. He was telling stories on radio and television even after he turned 80.

These stories suggest that mental and emotional problems can be treated as long as the causes leading to the illness are identified and removed. *The Yellow Emperor's Canon of Internal Medicine* already realizes that human beings fall sick because of a disorder of *qi*. Sometimes, that disorder can be corrected through diverting the attention of the patient or transforming the emotional state. The book also recorded that spasms can be treated by scaring the patient out of his *qi* disorder. After the patient regains composure, and his *qi* is back to normal, the spasm will be gone.

# Chapter 28

## Dispelling Suspicions

Paranoia is a common mental disorder. Many practitioners of TCM see paranoia as one of the causes inducing various kinds of sickness. They believe that when falling sick, people tend to dredge up all kinds of suspicions and fantasies. Some even die of prolonged suspicion or obsession. Others who would otherwise be perfectly healthy become seriously sick under the impact of suspicions.

A medical record kept by Xu Lingtai documents the story of a learned man named Li Minggu. Born into a poor family, Li was frustrated and mysteriously fell sick. He seemed to hear curses thrown at him every night, but he never saw who was cursing him. Meanwhile, the curses only got more rancorous. Annoyed, he could not eat or sleep and became restless. His uncle took him to a doctor. But when the doctor offered his diagnosis, Li snapped back, "I'm fine; it's only because someone is cursing me all the time." The doctor told him that was exactly the problem. His uncle assured him that nobody would criticise him, since his academic record and character were flawless. Suddenly, Li began to cry out aloud. "It's OK if it were somebody else trying to console me; but uncle, you knew precisely that I'm sick, and here you are, saying that I'm perfectly healthy. This doesn't make any sense. The man cursing me all day long yesterday next door must be you." His uncle explained that the day before he had been somewhere else rather than in his neighbor's home. "By the way, I don't even know who your neighbor is, how can I get into his home?" his uncle asked. The more he explained, however, the more suspicious Li became. Eventually, Li died.

*The Yellow Emperor's Canon of Internal Medicine* says that distress hurts the lungs, which, if not cured immediately, will spread to other internal organs and eventually affect the operation of *qi* inside the body. Paranoia or other self-induced deep suspicions, therefore, may well lead to death. Such sickness is normally beyond the reach of medicines and ordinary psychological therapy. The true therapy lies in the ancient tactic of deception.

Sun Zi contended that whether or not one is good at deception is one of the key factors for victory. In fact, ancient Chinese doctors had already noted that the tactic of deception had major medical applications. Zhang Zhongjing, a noted doctor in the Han Dynasty, proposed in his *Shang Han Lun* (*Treatise on Fevers*) the therapy of "treating deception with deception". A later doctor explained the therapy as: if somebody "rises without being roused, looking around in perfect health and mood; prevaricates when asked about what's going on and shows no pain; makes no sound when his pulse is being taken and his pulse is normal — all these are signs that the patient is faking sickness. The best way to cure such patients is to frighten them with threats of applying poisonous medicines and needles to them. Fright will dispel the fear."

*Trifling Words of Northern Dream*, a medical work by Sun Guangxian, recorded the following case. During the Tang Dynasty, a woman went with her husband to visit relatives. On the way, she mistakenly swallowed an insect. From then on, she became upset and fell sick. Many doctors were sent for, but none could treat her. Finally, a doctor named Yuan Zhen came. After learning of how the woman fell sick, he told one of the maids, "I'm going to administer to her medicines for diarrhea. When she vomits, you have to catch it with a vessel and tell her that you see a small worm swimming in it. Be sure that she doesn't find out the truth." The maid did what she was told, and the patient was so happy. Not long thereafter, she was back to normal.

Such cases are not rare, even in contemporary times. It was reported, for example, that a middle-aged woman began to find it difficult to swallow hard food after attending the funeral of a friend. Later, it became worse — she couldn't even drink water and lost a lot of weight. Finally, a professor came to her aid. The professor did some investigation, and it turned out that the woman's friend had died of esophagus cancer. Fearing that she herself might have the same problem, the woman eventually fell sick. The professor told her seriously, "We have a super-effective drug especially for this kind of ailment, and many have been cured. However, it hurts when injected." He told her family to get some delicious food for her. The patient felt pain with the injection and was deeply convinced. Once at home, she immediately did what she was told — eating a bowl of porridge and some fruit. To her surprise, it was no longer difficult to swallow. After a few more days of therapy, she was back to normal. She was so grateful to the professor that she paid a visit to his residence. "What drug did you use that was so effective?" she asked. "Distilled water," the professor answered.

These cases suggest that the doctors achieved therapeutical effects by "lying" to their patients, which served to dispel their unfounded suspicions. Hence the saying, "God will forgive doctors who lie to their patients."

It is not unusual to see people, especially youths, ignore their doctor's advice and refuse to undergo therapy or, when forced to, participate uncooperatively. In these circumstances, doctors are advised to use harsh rhetoric and even scary words to frighten them into undergoing therapy, a practice that is permitted by medical ethics. Again, *The Yellow Emperor's Canon of Internal Medicine* recorded that once a patient was very uncooperative during an acupuncture therapy, resulting in a disorder of his *qi* and blood. To solicit attention and cooperation, the doctor produced a long needle and showed it to the patient, warning, "I'm going to put it in real deep." Scared, the patient changed his attitude at once. The real depth of puncture, of course, depends on actual conditions.

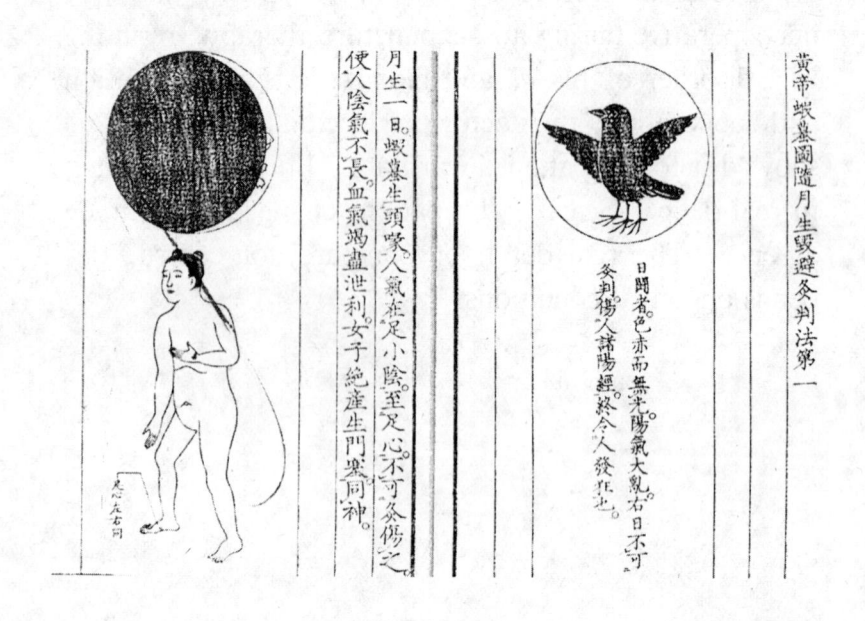

当日中见斗时(太阳黑子时),人体气血大乱,当禁止针灸 《黄帝虾蟆经》

Acupuncture is forbidden at mid-day, when the *qi* and blood of the body are running irregularly. (*The Yellow Emperor's Canon of Frogs*).

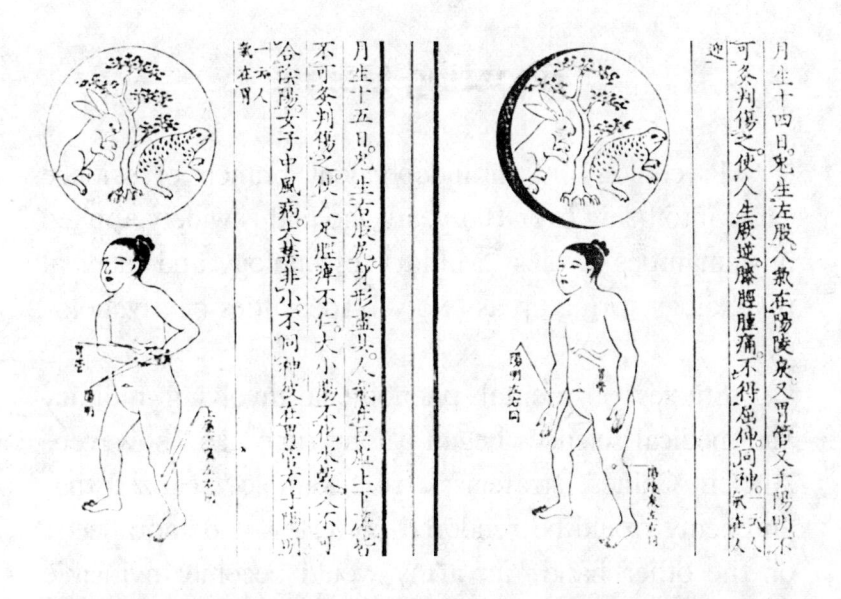

月生十四、十五日，人身的神和气随月亮的生(由缺向圆)和毁(由圆到缺)而游行於不同的经脉，针灸当避开神气所在之处。

日全食、台风暴雨、电闪雷鸣、暴冷骤热时，人体气机紊乱，此时禁针灸。

On the 14th and 15th of every lunar month, the essence and *qi* of the body run through different channels and collaterals in accordance with the wax and wane of the moon. Acupuncture should avoid these channels and collaterals.

When there is a total eclipse of the sun, a tornado, a storm or a drastic rise or fall in temperature, the flow of the *qi* of the body becomes irregular, and acupuncture is forbidden.

## Chapter 29

## Boosting Morale

Psychology as an independent branch of science came into being over 100 years ago and is widely applied in humanities studies. Military psychology and medical psychology spring up as two sub-specialties of psychology.

Studies of mental phenomena involving military and medical sciences began a long time ago, however. Ancient Chinese strategists, for example, realized that an enemy would be rendered powerless if demoralized; on the other hand, an army would become invincible with boosted morale. Demoralizing the enemy and boosting the morale of one's own army, therefore, became two basic components of psychological warfare.

Like so many other military strategies, this tactic has also found medical applications. *The Yellow Emperor's Canon of Internal Medicine* notes, "It is human nature to fear death and yearn for life. Patients will, without doubt, listen to their doctor if told frankly the nature of the sickness; are assured of recovery as long as they cooperate; are advised how to keep fit; and relieved of any tensions, fears and negative thoughts they might have." Basically, this means doctors preparing the patients psychologically for therapy by informing them of both the favorable and unfavorable conditions. This has been cited as the earliest reference to persuasive therapy in the history of TCM.

Long Qirui, a doctor in the Qing Dynasty, recorded a case involving persuasive therapy. One day, Long went to see a patient who ordinarily ate a lot, but suffered from prolonged diarrhea. Medication failed to take any effect for three months. To alleviate pain, he remained either seated or in bed every day. To his chagrin, it only exacerbated the diarrhea. Doctor Long, after making a diagnosis, assured him that the diarrhea would be gone in no time so long as he adjusted his diet and avoided sudden changes in temperature. "What worries me, though, is that you lock yourself in distress every day without doing any work," he told the patient. "You will eventually fall sick because the spirit is the owner of the body — when the spirit is weakened, the body falls sick. A laborer will not fall off even sleeping on the edge of a cliff because he fully concentrates his attention. A little boy is not scared at seeing a tiger, but instead will chase the animal with a stick because he does not know that tigers can be harmful. You are actually not sick, and you feel sick only because you believe so. You'll be fine if you get rid of those groundless fears and enjoy life."

The patient, thinking that what the doctor told him made sense, took the doctor's advice and in less than three days, he was back to normal.

Modern medicine indicates that persuasive therapy is useful for alleviating pains suffered by cancer patients who are aware of their illness. The doctor must first of all listen to the patient's account of his or her illness, psychological experience and desires in a sympathetic, caring and amicable manner. They should never arrive at a rash conclusion, interrupt the flow of conversation, initiate persuasion prematurely, or show impatience. This gentle approach is not only helpful for gaining a thorough understanding of the mental world and personality of the patient — which will lay the foundation for later therapy — but will also help relieve the patient of their protracted depression, which will alleviate sickness. On the basis of that, the doctor then offers his or her account of the patient's conditions and assurances of therapy. The main thrust of persuasion should be placed on projecting a positive prospect for the patient and helping them build up confidence in therapy. Repeated assurances of the success of therapy, reinforced by effective medication, are conducive to fostering positive thinking and promoting recuperation. This is important especially for neurotic and suicidal types of patients.

## Chapter 30

## Guidance

In modern warfare, an army on the move is greatly aided by sophisticated navigation equipment and maps. In ancient times, however, troops often had to rely on local guides for passage through a strange land.

Sun Zi said, "Unless you employ local guides, you cannot turn the terrain to your advantage." Favorable climatic, geographical and human conditions are the three keys to victory and to anything for that matter. None of the three factors are disposable. Local guides are especially important in ensuring the right direction for an army. Xu Dachun, a noted physician in the Qing Dynasty, believed that in medicine, that same principle applies — application of medicine must be guided. He proposed a formula to "apply medicine according to the meridians and collaterals".

Ancient physicians discovered in their practice that each medicine applies to a certain illness. For example, although all cold-natured medicines can clear away heat, they each actually apply a particular kind of heat: some are good for clearing away heat from the lungs, some good for the liver, and still others for the stomach. The same is true of invigoratives. Some are good for the lungs, some for the spleen, and others for the kidney. Traditional Chinese pharmacopeia, based on the theory of differential diagnosis according to the state of viscera and the theory of meridians and collaterals, assigns all the medicines to 12 meridians. This is known as the meridian tropism of medicines.

The meridians and collaterals are like the guides — misguidance can lead to a fiasco in a battle and the danger of mistreatment in medicine. Perhaps a little more explanation of the meridians and collaterals is needed here.

The meridians and collaterals, both unique terms used by TCM, refer to the channels of the *qi* and blood all over the body connecting various organs. The meridians are the main channels, and the collaterals network-like branches. Most of the meridians run deep inside the human body, while the collaterals circle on the surface. In addition, meridians follow a fixed path, while the collaterals crisscross all over the body, connecting all the viscera, organs, orifices, skin, muscles and bones. The meridians and collaterals are interconnected directly or indirectly, constituting a complete system.

In this organic whole, the most predominant channels are the 12 meridians, the main paths for the circulation of *qi* and blood. They are evenly distributed on the front, back and both sides of the human body, circulating along the inner or outer sides of the upper or lower limbs. Each meridian corresponds to a viscera. The name of these 12 meridians consists of three parts; hand or foot, *yin* or *yang*, solid or hollow organs. Hand meridians run from the upper limb to the visceras, while foot meridians run from the lower limbs to the visceras. Likewise, the *yin* meridians that run along the inner side of the limbs correspond to solid organs, while the *yang* meridians that run along the outer side of the limbs correspond to the hollow organs. The names of the 12 meridians are as follows:

The three *yin* channels of the hand: the lung channel, the pericardium channel, and the heart channel;

The three *yang* channels of the hand: the large intestine channel, the three-*jiao* channel, and small intestine channel;

The three *yin* channels of the foot: the spleen channel, the liver channel, and the kidney channel; and

The three *yang* channels of the foot: the stomach channel, the gallbladder channel, and the urinary bladder channel.

Because the meridians and collaterals connect the internal and external parts of the body, a surface ailment can affect the visceras and vice versa. These channels, therefore, serve as indicators of disease affecting various parts of the body. When the lung channel is diseased, for example, the symptoms usually are coughing and asthma; when the liver channel is morbid, the main manifestations are hypochondriac pains and spasms; when the heart channel is affected, palpitation and insomnia will ensue.

Because different medicines by nature treat different diseases, they are said to belong to the different corresponding channels. Balloonflower root and bitter apricot seeds, for example, are good for alleviating coughing and an oppressed feeling in the chest and therefore belong to the lung channel; the antelope's horn and hooked uncaria are used to check hyperactivity of the liver and relieve convulsion and therefore belong to the liver channel; and cinnabar and poria are used to tranquilize and relieve mental stress, they are said to belong to the heart channel.

Some drugs belong to more than one channel, however, indicating that they are curative of several kinds of diseases. Gypsum, for example, belongs to both the lung and stomach channels, suggesting that it is used to remove intense heat of both the lungs and the stomach. Pilose asiabell root, similarly, belongs to both the spleen and lung channels, indicating that it is capable of invigorating the vigor energy of both organs.

Sometimes a morbid channel or collateral can affect others and therefore requires a combination of several kinds of drugs. For example, pneumonia might be compounded by a weak spleen, and as a result, therapy requires medicines for both the lungs and the spleen — spleen can promote recuperation of the lungs. It is imperative, therefore, not to be rigid about matching medicines with channels clinically.

The channels thus serve as "local guides" in therapy. A heated liver and red eyes indicate that those drugs that are good for removing the heat from the liver should be used. On the other hand, a cold stomach and pains in the abdomen dictate the application of cold-dispelling medicines that at the same time warm the stomach and alleviate pains. It is very important for doctors to understand the correlationships between major medicines and various channels.

Knowledge of those correlations will not only enable doctors to prescribe the right medicines for the right disease, but will also facilitate the prescription of a "conductant drug", a medicine that directs the action of a prescription to the affected channel. Balloonflower root, for example, directs the action into the lungs; cinnamon bark directs the action into the kidney; cyathula root directs the action into the lower limbs; mulberry twigs direct the action into the upper limbs; bugbane rhizome directs the action upward; and eagle wood directs the action downward.

手阳明大肠经穴图/起于食指之端,沿手臂外侧前缘上行至肩前,共二十个穴位　清代黄谷绘

The Large-Intestine Channel of Hand-*Yangming*, which begins at the tip of the index and runs to the shoulder along the outer side of the arm, encompasses 20 acupoints. (Painted by Fei Gu of the Qing Dynasty)

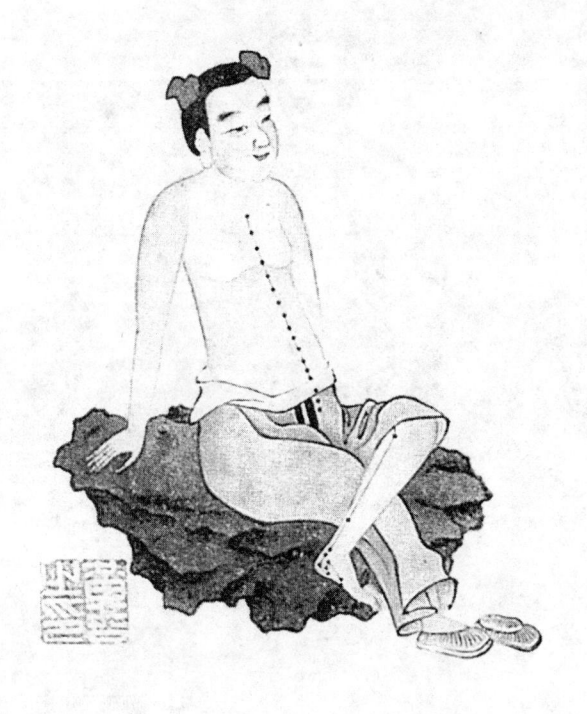

足少阴肾经穴图/起于小趾至舌根共二十七个穴位,主治泌尿、生殖以及呼吸、消化等病症。

The Kidney Channel of Foot-*Shaoyin*, which begins at the small toe and ends at the base of the tongue, comprises 27 acupoints. This channel is said to be associated with the urinary, reproductive, respiratory and digestive systems.

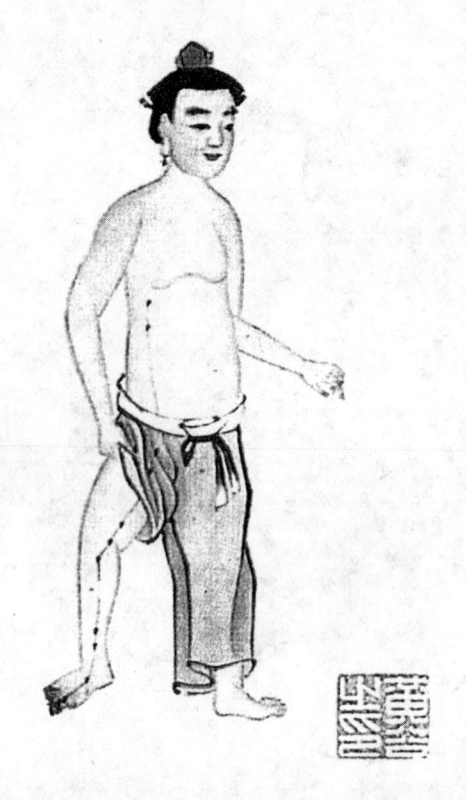

足厥阴肝经穴图/起于大趾，上行胫骨内缘，沿咽上行至巅顶，共十四个穴位，主治泌尿生殖、神经系统、肝胆病症。

The Liver Channel of Foot-*Jueyin*, which begins at the big toe and runs along the inner side of the ribs and the throat to the top of the head, consists of 14 acupoints. This channel is said to be associated with the urinary, reproductive and nervous systems and the liver and gallbladder.

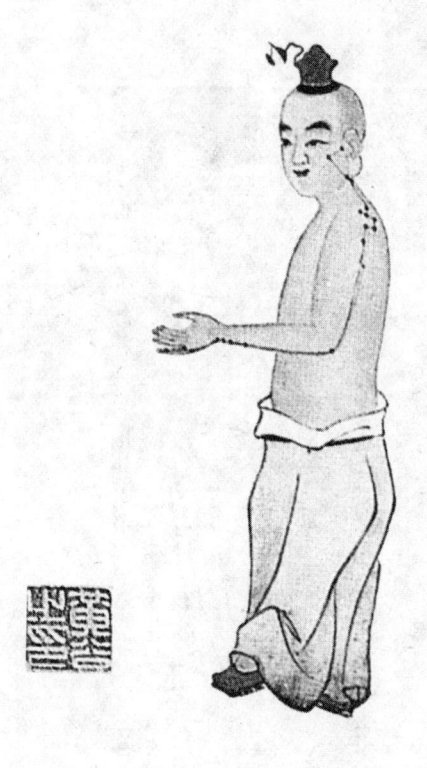

手太阳小肠经穴图/起于小指之端,沿手臂外侧上行,共十九个穴位。

The Small-Intestine Channel of Hand-*Taiyang*, which begins at the tip of the little finger and runs along the outer side of the arm, encompasses 19 acupoints.

手少阳三焦经穴图/共二十三个穴位,主治胸、心、肺、咽喉等病症。
The Tri-*Jiao* Channel of Hand-*Shaoyang*, composed of 23 acupoints and said to be associated with the chest, heart, lungs and throat.

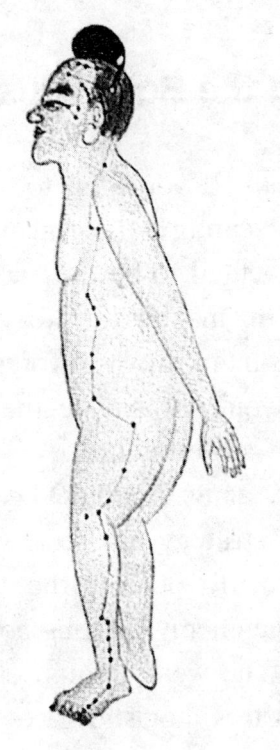

足少阳胆经穴图/起于外眼角,入第四趾端,共四十四个穴位,主治胸胁肝胆、热性病症、五官病症等。

The Gallbladder Channel of Foot-*Shaoyang*, which begins at the outer corner of the eye and ends at the tip of the fourth toe, comprises 44 acupoints. This channel is associated with the chest, liver, gallbladder and the five senses and heat syndromes.

# Chapter 31

# Removing the Root Causes

As the saying goes, "If you want to round up bandits, you have to first capture the ring leader." *The Thirty-Six Strategies*, reputed to be the "non-official art of war", agrees with this maxim. Strategy 18 says, "If you destroy an enemy's main thrust of force and capture its commander, then you will subdue the entire army like stranding an out-of-sea dragon."

Hitting the enemy at its Achilles' heel has always been a major military strategy advised throughout the centuries. Sun Zi said, "To operate the army successfully, we must avoid the enemy's strong points and seek out his weak points." The weak points refer mainly to the command headquarters, logistic bases and key defense posts. A well-known battle in the Tang Dynasty — Li Su Raids Caizhou on a Snowy Night — well illustrates the point.

Once the Tang Emperor Xianzong dispatched troops to attack Wu Yuanji, a self-appointed pretender, in Caizhou (modern-day Runan, Henan Province). The attack was thwarted, however, and a stalemate ensued. In the winter of 816, General Li Su, after learning that the defense in Caizhou was very weak, decided to attack the enemy's headquarters head-on. To secure victory, he sent an advance contingent to assault Wufang, a town in the neighborhood of Caizhou. After wiping out 1,000 enemy troops, the contingent returned to their camp. This move tested the enemy's defenses without infuriating the enemy.

On the night of October 15, a northerly wind was raging ferociously, and snowflakes were dancing across the land. Li led 5,000 elite troops to attack Caizhou. Although the headquarters of the enemy troops, the city had not seen war for over 30 years. The storm further relaxed the guard of the defenders. By daybreak the Tang army had arrived at the foot of the city wall and began climbing the walls. Wu Yuanji woke up from a dream and hastily threw up some resistance. It was too late, however, for the Tang army had already taken up positions everywhere, and there was no way of pushing them back. Wu surrendered.

Xu Dachun, the noted physician in the 18th century, proposed in his *Treatise on the Origin and Development of Medicine* that in medicine, just like in a battle, it is very important to attack the enemy's headquarters head-on.

The most important therapeutic principle in TCM is to remove the root causes of illnesses. The "root" causes differ from the "superficial" causes. Any ailment develops out of primary and secondary causes. The "root" causes are the primary ones, and the "superficial" ones are secondary. These two concepts can be used to illustrate many situations. The relationship between body resistance and pathogenic factors, for example, can be explained as one between the primary and secondary causes. So is the relationship between the origin and manifestation of an illness. Likewise, the internal organs are primary, and superficial parts of the body are secondary; the original illness is primary, and the new one secondary.

Treatment is possible only after the primary causes are identified. TCM holds that all illnesses must manifest themselves through some symptoms; the symptoms, however, are only superficial phenomena rather than the essence of the ailment. Only after identifying the root causes can doctors prescribe the right remedy and cure the disease. A patient with abundant expectoration, for example, need not be treated with medicines to purge phlegm; rather, because the spleen is the source for phlegm, the fundamental resolution lies in invigorating the spleen.

Likewise, bleeding is essentially related to the functioning of *qi*; anhydrosis (lack of sweat) is tied with the heart; fevers are caused by weak *yin* forces; dyspnea can be alleviated through improving the kidneys; seminal emissions result from an affected organ.

Having said this, however, it must be pointed out that the authors have no intention of advising doctors to ignore the symptoms. The point is for doctors to distinguish between the primary and secondary causes of illnesses and apply the corresponding medicines.

Depending on the nature of a disease, the doctor should exercise some flexibility in applying this principle. Generally, efforts should be focused on removing the primary causes; sometimes, however, it is necessary to take into account both primary and secondary causes; still, in some critical instances, the doctor may start with the secondary causes and then move on to tackle the primary ones.

Cao Yingfu, an outstanding modern physician, once recorded the following case in his medical files. A patient came to him complaining of pains in the loins. An examination revealed that his right testicle was bigger than his left one, and there was a bun-shaped swelling to the left of his abdomen. When eating, his abdomen first swelled and then hurt acutely. An hour later, he would feel a bubbling noise in his abdomen, and the pains would subside. Unable to sustain the excruciating pains, the patient attempted to commit suicide several times. Dr. Cao diagnosed that the pains resulted from an invasion of cold-dampness that lingered in the liver, causing the *qi* and blood to stagnate. Therefore, to remove the pains, the liver and kidneys needed to be warmed and invigorated so as to activate the *qi* and dispel the cold-dampness. The prescription worked and the patient recovered his health fairly quickly.

Chapter 32

## Bolstering Resistance

As the saying goes, "When a disease strikes, it does so like a towering mountain falling down." Some serious ailments develop rapidly and momentously. A "speedy" treatment with powerful medication is not helpful, but on the contrary will aggravate the ailment. Xun Dachun contended that the appropriate approach is to "weaken the enemy".

"Weakening the enemy" is originally a military strategy designed to wear an enemy down so that it loses its combat capability. Fan Li, a noted strategist in the late Spring and Autumn Period, argued that a wise general must be good at wearing an enemy down, while covertly boosting his own force.

According to Xu, when an illness first strikes, it is unnecessary to pay too much attention to it; rather, the right thing to do is to preserve the vital energy and "wear down the enemy". In TCM, this is known as "strengthening the body resistance to eliminate pathogenic factors".

According to practitioners of TCM, the development of a disease is in large measure a process in which the body resistance fights the pathogenic factors. If these pathogenic factors prevail, then an illness arises; otherwise, the illness subsides. The therapeutic process, therefore, is an effort to strengthen the body resistance and eliminate pathogenic factors.

Body resistance can be reinforced through a variety of methods, such as applying medicines, improving nutrition and doing exercises to strengthen the functioning of the *qi*, blood, *yin* or *yang* forces.

On the other hand, pathogenic factors can be dispelled through applying medication, acupuncture or surgery. Clinically, this includes such therapies as inducing diaphoresis, purgation, removing food retention and promoting digestion, removing blood stasis, causing vomiting, and removing dampness and wind.

Depending on the nature of the disease, strengthening body resistance and removing pathogenic factors may be applied separately or in conjunction clinically. If body resistance is weak and the pathogenic factors are active, then the two therapies must be applied simultaneously — usually in the form of using invigoratives and purgatives at the same time. The principle is that body resistance must be reinforced without aiding the pathogenic factors; pathogenic factors must be removed without harming body resistance. Take the following case as an example.

According to the *Cases of Famous Physicians*

*Ancient and Modern*, once Zhu Danxi, a noted physician in the Yuan Dynasty (1271-1368), received a senile patient. The patient, having worked long hours outdoors in the cold weather, had headaches and fevers, compounded with arthritis, anhydrosis and unconsciousness. In the early stages of the onslaught, the patient took some Ginseng and Perilla Drink to dispel the cold. After the sweating became profuse, however, the patient started to run a high fever. Dr. Zhu diagnosed that the patient suffered from internal injuries — a weak stomach resulting from hunger, coupled with hard labor and cold weather, leading to the exogenous pathogenic cold invading the body. He then proceeded to prescribe blood-reinforcing medicines such as ginseng, astragalus root, root of Chinese Angelica, rhizome of largehead atractylodes, tangerine peels and liquorice root, combined with prepared aconite root and peony decoction. He told the patient to take five doses per day. After 10 days, the patient felt much better. The doctor then added gravy and desert cistanche porridge to improve his appetite. After another 10 days, the patient reported that the symptoms were gone.

The key to therapy in this case was that the exogenous cold was just a superficial cause, and the real cause was internal injury. Because of the senility of the patient, it was inappropriate to induce diaphoresis for fear of damaging bodily resistance. Therefore, Dr. Zhu resorted to invigorating the *qi* and blood first before dispelling the exogenous cold from the body. He then added prepared aconite root to tone up the *yang* elements and peony decoction to tone up the *yin* elements. As a result, the stomach *qi* became full, and the body began to sweat naturally, constantly weakening the exogenous cold within the body. Eventually, the cold was discharged along with diaphoresis.

Therefore, the military tactic of "wearing the enemy out" is analogous with the medical principle of "strengthening body resistance to eliminate pathogenic factors". It is, indeed, the most fundamental and effective therapy.

## Chapter 33

# Preventing Disease

As the great Qing Dynasty physician Xu Dachun pointed out in his *Treatise on the Origin and Development of Medicine*, "Pathogenic factors invading the body must be stopped early, the way an enemy's advance must be blocked." He was talking about one of the most important principles in TCM: applying therapy early in the stage of development so as to prevent proliferation of pathogenic factors.

Proponents of the exogenous febrile disease theories believe that pathogenic factors invade the body through the surface and via the six channels — *Taiyang*, *Yangming*, *Shaoyang*, *Taiyin*, *Shaoyin*, and *Jueyin*. These are originally names of the 12 meridians for the limbs. Zhang Zhongjing, a physician in the Han Dynasty, based his exogenous febrile disease theory on the 12 meridians. He divided the invading pathogenic factors into six categories according to the location of the affected collateral/meridian, viscera or body part and the degree of affection. He named these the "six-channel diseases". The development of these six diseases is actually the process of the pathogenic factors invading the body from the surface to the inside. A *Taiyang* disease is the initial stage where pathogenic factors are still embryonic; a *Yangming* disease is the final stage where the pathogenic factors are in full bloom; a *Shaoyang* disease is the transitional stage where the pathogenic factors are halfway between the surface and the inside of the body. Weakened body resistance can result in a *Sanyin* disease characterized by a feeble spleen; a *Shaoyin* disease is characterized by weakened heart and kidney functions, while a *Jueyin* disease is marked by alternating cold and heat.

The doctrine of epidemic febrile disease, in addition, holds that pathogenic factors invade the body primarily through the mouth and the nose and thereby undergo different stages of development.

As discussed above, most of these pathogenic factors invade the body progressively, from the surface to the inside and from the external to the internal. Therefore, *The Yellow Emperor's Canon of Internal Medicine* advises that the best doctor follows an order of therapy beginning with the skin, then the flesh, then the channels, and finally the viceras. The key is to apply therapy before it's too late.

The parallel between a pathogenic factor invading the body through a channel and an enemy making an inroad is obvious. Militarily, to keep an enemy at bay, it is necessary to control the strategic positions and wipe out the enemy before he penetrates deep into one's territory. As Sun Zi pointed out, "Those who are able to keep the enemy from coming do so by creating obstacles and inflicting damage on him."

Wang Haogu, a physician in the Yuan Dynasty, observed, "Militarily, if the enemy is in a valley, block the mouth; if the enemy is on a riverside location, control the ferry; if the enemy is on a flatland, clear 1,000 *li* (a *li* equals 500 meters) of land. Medically, blocking the springs is analogous to putting a needle in the *yu* acupoint; controlling the ferry to subduing a disease in its embryonic stage; and clearing 1,000 *li* of land to taking large doses of medicine to foster the body's resistance."

The idea of stamping out a disease in the early stage of development can be traced back to *The Yellow Emperor's Canon of Internal Medicine*. When a certain internal organ is to become morbid, a purplish redness will appear on the corresponding part of the face. Applying acupuncture at this stage can effectively prevent the affection of the organ. For example, redness on the left cheek suggests heat in the liver, while redness on the right cheek indicates heat in the lungs, according to the medical classic.

## Chapter 34

## Peace of Mind

TCM holds that the human body is an organic whole, with the most important parts being the 12 internal organs. *The Yellow Emperor's Canon of Internal Medicine* illustrates the functions and interactions between the 12 organs.

Once the Yellow Emperor asked Qi Bo, the master physician, about the interactions between the 12 internal organs. Qi explained the interrelationships as follows.

"Of all the internal organs, the heart is like the monarch, from which man's wisdom stems; the lungs are like the prime minister, governing the *qi* of the body and regulating the body movements; the liver is like a general who makes all the strategies; the gallbladder is capable of making decisions; the sternum is like a minister who conveys the emotions of the monarch; the stomach is like a warehouse that stores all the nutrients the body needs; the large intestine is like a pipeline that digests, absorbs and discharges food; the small intestine further decomposes the food already digested by the stomach; the kidneys are the source of energy, wisdom and skills; the triple energizers manage the bodily fluids; and the urinary bladder is where the fluids converge and urine is discharged.

"Together, these 12 internal organs are interrelated. The monarch, of course, is the most important. Following this principle in preserving health, one will live a long life. Applying this principle in governing a country, the country will be prosperous. Violation of the principle, on the other hand, will damage the health and put the nation at risk."

These comments were made more than 2,000 years ago. They suggest that the 12 internal organs are interconnected and interactive. If one of them malfunctions, the others are inevitably affected; on the other hand, if the other organs are reinforced against the sickness, then pathogenic factors will be checked from spreading to other body parts.

Great physicians over the dynasties had paid much attention to the principle of preventive treatment. Prevention refers to both protecting the body from pathogenic factors and protecting internal organs from being affected once those factors have invaded the body. It is therefore a combination of prevention and treatment. *Classic on Medical Problems*, a masterpiece of TCM by an anonymous physician, points out that the best doctor will reinforce the resistance, through applying tonics and invigoratives, of relevant internal organs before they are affected by a morbid organ. This is necessary for checking the proliferation of the illness. The mediocre doctor treats a morbid organ for its own sake without taking into account its implications to other related parts of the body.

The famous Qing Dynasty physician Xu Dachun proposed, "The best way to prevent a raging disease from spreading to major parts of the body is to defend those key strategic positions susceptible to enemy attacks." This is a major example of introducing a military strategy into therapy.

## Chapter 35

## The Roots of Disease

As the old adage goes, grain and fodder must be prepared in advance for the troops and horses. Sun Zi noted, "An army deprived of equipment, food and fodder and material reserves will perish." Therefore, a wise commander will incapacitate an enemy by destroying its logistic supplies with fire.

During the Three Kingdoms Period, Cao Cao and Yuan Shao, two contenders for power, were locked in confrontation. Yuan, however, enjoyed a cutting edge over Cao because his troops outnumbered Cao's at least 10 to one. Cao eventually beat Yuan, however, because he burned a logistic supply base of Yuan's troops.

The Qing Dynasty physician Xu Dachun believed that this strategy had medical applications. For example, he pointed out in his famous *Treatise on the Origin and Development of Medicine*, the best treatment for diseases arising from gluttony is to "burn the food and fodder supply of the enemy", or eliminate the root causes of the disease. In the case of gluttony, the therapy is to discharge the undigested food in the stomach or the virus-carrying food through applying emetics. The following case in the Yuan Dynasty illustrates the point.

One day, a middle-aged man went out hunting with friends and brought down a few wild rabbits. They built a fire and roasted the rabbits. Everyone got one piece, but the middle-aged man got an extra one. By the time he got home at dusk, he felt thirsty and downed several bowls of milk. That evening, the man had a swollen stomach and excruciating pains. He couldn't sleep or sit straight. Nor could he vomit or excrete. His family sent for Luo Tianyi, the famous doctor of the time. After diagnosis, Luo concluded that the patient suffered from excessive food and drink, which damaged the intestines and the stomach. He explained that too much roasted meat causes thirst; and once diluted with milk, it fills the stomach and causes pains.

He then prescribed emetics to induce excretion and vomiting. Three days later, he added thin porridge and ginseng tonics. In less than seven days, the patient recovered fully.

The above case suggests that the best therapy is to eliminate the root causes of diseases, be it external or internal. The following case provides another example of "burning the food and fodder supply of the enemy".

Juanjuan, a 15-year-old only child, was physically so weak that she chronically caught cold characterized by headache, sneezing, fever, fatigue and perspiration, despite the meticulous care of her parents and grandparents. Doctors were divided in their diagnosis. Some suggested that it's just the common cold which is no cause for concern, and others believed that over time the disease would cure itself. Several years passed, however, and the girl was still weak.

One day, her great uncle, an old practitioner of TCM, visited the family. After inspecting the girl's countenance and the environment, the physician blamed the parents and grandparents for shutting the windows and keeping the girl too warm. "The girl catches cold so often because she is like a flower in a greenhouse, isolated from nature." He continued that the girl suffered from none other than a deficiency syndrome, a weak body resistance, which allowed invasion by bacteria and viruses.

He then prescribed a medicine consisting of astragalus root, bighead atractylodes rhizome and ledebouriella root to replenish the *qi*, reinforce the spleen, induce diaphoresis and dispel cold. He also emphasized the importance of physical exercise, especially outdoor activities. Such exercises, he said, would build up the body's resistance against pathogenic factors. Following his instructions, the girl finally recovered.

## Chapter 36

# Holistic Treatment

It is not unusual for an elderly person to catch cold, while he or she already suffers from chronic bronchitis. In TCM, the original illness is called "*ben*", or the primary illness, and the ensuing sickness is called "*biao*", or secondary illness. Xu Dachun likened the original sickness to the agent provocateur of the enemy, and the ensuing sickness to a frontal assault. To treat a compound disease, it is necessary to "eliminate the agent provocateur", he said.

In war, when threatened by both a frontal enemy and a supporting force from inside, the right strategy is to focus on the frontal enemy, while using a small force to check the supporting force. Xu applied this strategy in medicine, suggesting a holistic approach to compound diseases.

There is no fixed formula for treating such diseases, though. Sometimes it is necessary to treat the primary disorder; at other times it is necessary to treat the secondary ailments; still, sometimes it is necessary to treat both the primary and secondary diseases at the same time. For example, patients with constipation, body fever and dry lips — all symptoms of pathogenic heat syndrome — should be treated with both therapies for primary and secondary diseases simultaneously. That is, using purgation to relieve the internal heat as treatment for the primary symptoms and nourishing the *yin* elements as treatment for the secondary symptoms. If only purgation is used, the body fluids, which are already being dried up by the heat, might be further exhausted and damaged; on the other hand, if only nourishment is used, then the heat cannot be relieved.

The same principle is true of diarrhea, whose main symptom is tenesmus, with bloody excreta, a yellowish tongue and weak pulse. Because damp-heat is the primary cause, treatment should focus on removing the damp-heat first. However, this is not enough; the secondary causes of abdominal pains and tenesmus must also be treated.

In reality, the primary and secondary are interchangeable sometimes. Therefore, flexibility is needed in clinical practice. As mentioned above, sometimes it is necessary to focus on the primary symptoms, while giving consideration to the secondary ones. At other times, however, it is necessary to focus on the secondary symptoms, while paying attention to the primary ones. Feilao, or impairment of the lung by overstrain, is characterized by a deficiency of the lung-*yin* (primary symptom), with secondary symptoms such as afternoon fever, night sweats and coughing. Therapeutically, the focus should be placed on the primary symptom, while attending the secondary symptoms. However, if hemoptysis occurs, then the above therapy should be reversed.

The key to therapy, therefore, lies in the ability to distinguish between the primary and secondary symptoms.

# Chapter 37

## Understanding the Six Channels

Ke Qin, an eminent physician in the 18th century, compared the "six channels" — *Taiyin*, *Shaoyin*, *Jueyin*, *Yangming*, *Taiyang* and *Shaoyang* — with terrain in war. As Sun Zi said, "Advantageous terrain can be an ally in battle." Ke argued that understanding the six channels can be a key to treating all kinds of diseases.

The six-channel theory is based on an analysis of the symptoms accompanying an exopathogenic disease with regard to the characteristics and locations of the morbid condition, the balance of pathogenic factors and body resistance. According to Ke, all diseases coming through the six channels are either of *yin* or *yang* origins. Those from the three *yang* channels are mostly caused by exopathogenic factors whose symptoms can be diagnosed on the skin; and those from the three *yin* channels are mostly caused by internal pathogenic factors whose symptoms can be diagnosed only through the five viscera (heart, liver, spleen, lung and kidney).

It follows, therefore, that exopathogenic diseases should be treated with medication that removes such factors from the body, such as ephedra decoction for removing the cold coming from the *Taiyang* channel, and minor decoction of bupleurum for allaying the fire coming from the *Shaoyang* channel. On the other hand, diseases caused by internal pathogenic factors associated with the three *yin* channels should be treated with medicines that boost the body resistance. This includes ginseng decoction for tonifying the spleen-*yang* and dispelling cold, the Zhenwu decoction for strengthening the kidney-*yang* and removing retention of water, and Black Plum pill for curing diseases associated with the *Yueyin* channel.

# Chapter 38

## People and the Army

Sima Fa, an ancient military strategist, once remarked that what is effective for administering an army may not be applicable to administering the public and vice versa. This is because an army is managed through strict enforcement of rules and severe punishment of violators, whereas the public cannot be managed in such a way.

It is possible, however, to apply the military rules to medicine. Su Dongpo, a great poet in the 11th century, related a story in his diary. One day, the poet went on an excursion with his physician friend Zhang Wenqian. Suddenly, his eyes hurt, and he tried to ease the pain by cleaning them with hot water. His friend stopped him, saying, "When your eyes hurt, rest them; when your teeth ache, labor them. This is because the eyes are like the people, and the teeth the army."

This analogy makes good sense. It has been proven that clicking the teeth is a good way of maintaining dental health. In addition, the human eye is a very delicate organ equipped with a complete defensive system. Even a fine granule inside the eye can cause excruciating pain. One of the most important ways of protecting the eyes is "not to exhaust them excessively", just as the government treats the people. When something goes wrong with the eye, it is inappropriate to cleanse it with hot water because that will only stimulate it even more. The best way is to rest it and let the pain heal itself.

The difference between the eyes and the teeth is indeed like the difference between administering the people and an army.

# Chapter 39

# Different Approaches to One Disease

Xu Dachun, in his *Treatise on the Origin and Development of Medicine*, proposed that different approaches be adopted to treat the same disease. In explaining his theory, he cited a military strategy whereby a inferior force beats one with superior numbers.

Generally, the reserve is true. However, it is possible to turn the tables if flexible tactics are employed. As Sun Zi said, "If we're able to determine the enemy's disposition while concealing our own, then we can concentrate our forces while his are dispersed. And if our forces are concentrated at one place while his are scattered at ten places, then it is ten to one when we attack him at one place. This means we will be numerically superior. If we are able to use many to strike few at a selected point, those we deal with will be greatly reduced in number."

The key to beating a superior force is to get the enemy's forces scattered, while concentrating one's own at one place. Xu drew on this principle and proposed that a disease be "divided" and treated. According to Xu, different approaches should be adopted to treat the same disease for the different symptoms and during different stages of development. In this way, even a small dosage can be highly effective. The different symptoms of the disease, once divided and broken from each other, cannot affect or support each other as a retreating army.

For example, dysentery has different symptoms, such as passing stool with blood and pus, abdominal pains and tenesmus (an ineffectual and painful straining in defecation due to stagnation of *qi* resulting from the attack of damp-heat pathogen). In therapy, therefore, these symptoms should be treated with different methods. Promoting the flow of *qi* can remove tenesmus and abdominal pain, while promoting blood circulation can eliminate the blood and pus.

Measles, likewise, has different symptoms during its different stages. During the initial stage, the pathogenic factors affect primarily the exterior area, and therapy should feature disphoresis to promote eruption of the measles so as to prevent complications. The intermediate stage is characterized with scorching lung-heat, calling for medication that removes the heat and toxic. In the final stage, the remaining heat inside the body still hurts the lungs and stomach, and therapy should concentrate on nourishing the *yin* elements of the body and clearing away the residual heat. In this way, the disease is broken down into different sections that are isolated from each other. Eventually, unable to communicate with each other, they ebb away. This method is known in TCM as "treating the same disease with different methods".

This method also implies that different treatments be applied to the same disease depending on the nature of pathogenic factors, reactions of the body and the living environment of the patient. Asthma, for example, can be induced by cold, heat, weakness and pathogenic factors. In addition, every individual is different from each other in terms of physique, sex, age and living environment. These differences entail different treatments. Even for the same patient or patients of the same age and physical build, therapy varies from winter to summer, from a humid to a dry season.

Therefore, once the laws of the evolution of a disease are known, the "miracle" of victory by the inferior force can happen. The following case well illustrates the point.

A few years ago, there was a patient who had been suffering from asthma for over 20 years. He happened to live in north China where it was very cold in winter. The cold weather triggered his asthma, with phlegm resembling bubbles. A doctor diagnosed that the cold weather was responsible for the patient's asthma and prescribed Minor Decoction of Green Dragon. The patient took the medication, and soon his asthma was gone. The next year, he moved to the south. On his way, his asthma broke out again due to fatigue. He used the old prescription and cooked the same decoction. It failed to work, however, and his asthma only got worse. His complexion got pale, his body listless, and he sweated profusely — all symptoms of *qi* deficiency. After taking some tonics for invigorating *qi*, he recovered, and his asthma did not return for two months. One day, however, when he smelt gas and argyi leaf smoking, his asthma recurred. It was already autumn, and the weather was very dry. The patient used the same tonics for invigorating *qi* to no avail. When the doctor came, the patient had a reddish complexion, dry mouth and reddish tongue, accompanied by headaches. This indicated that the source of the asthma had changed from cold to heat. The doctor prescribed a cold-natured decoction of ephedra, apricot kernel, gypsum and licorice, immediately easing the coughing. Subsequently, the doctor, believing that the root cause for the asthma was weak kidneys, gave the patient some medicines to tone up the kidneys. This improved the physique of the patient fundamentally, his asthma did not recurring for years.

**Chapter** 40

# One Approach to Different Diseases

Contrary to the previous principle of treating the same disease with different approaches, the principle of using the same approach to different diseases, also proposed by Xu, posits that the same treatment can be applied to different morbid parts of the body, different functions of the viscera and different diseases as long as they stem from the same origin. For example, diarrhea, prolapse of rectum and prolapse of uterus in the same patient all result from an insufficiency of *qi* and therefore can be cured with medicines for invigorating the *qi* of middle-*jiao*. Once that is done, the various symptoms will be checked from developing further. They will also be isolated from each other and, like a retreating army, finally collapse one by one.

In TCM, this theory applies not only to different diseases sharing the same origin, but also to different diseases having the same symptoms. Take the following case of a patient named Xu. Because of indiscretion in the food he ate, he ran a fever for several days in a row, feeling dry in the mouth and tired. His urine was reddish, and he felt stagnant while urinating. Furthermore, his legs were sore and weak, pulse swollen and tongue pale. These are symptoms of deficiency of kidney-*yin* (essence and fluids), compounded by the flaming of the kidney fire. Correspondingly, therapy should focus on kidney deficiency and the flaming kidney fire, i.e. bolus of six drugs including rehmannia.

Another patient named Yao suffered from diarrhea with no appetite. All year round, he was listless, with a withered complexion. These are symptoms of deficiency of spleen- and kidney-*yang*. The doctor prescribed a decoction to tone up the patient's *qi* of middle-*jiao*. However, he took out the Chinese angelica root and added nutmeg instead to raise the sinking *yang qi*. In addition, he used ginger to produce the earth elements within the body. After four days, the diarrhea stopped, but the phlegm continued unabated. The doctor then used the eight-ingredient decoction together with a decoction for strengthening the middle-*jiao* and replenishing *qi*. Soon, the phlegm disappeared.

These two cases, though showing different symptoms and causes, share the same pathogenesis. Therefore, both were cured with the decoction for strengthening middle-*jiao* and replenishing *qi*, the only difference being that one or two ingredients of the decoction were adjusted depending on the circumstances.

Practice has shown that diversified as they may seem, all symptoms share something in common. This is because all diseases manifest themselves through the human body, and they must of necessity follow some kind of law. Therefore, for different diseases under the same category or at the same stage, the same therapy can be applied. For example, diarrhea, dysentery, irritable colon, bronchial asthma and chyluria all share the symptoms of listlessness, indolence, low voice, shortness of breath and profuse sweating. In these cases, some *qi*-strengthening medicines can be used. Still the decoction to strengthen the middle-*jiao* and replenish *qi* cannot necessarily cure all these diseases — after all it is not capable of rooting out the original causes of those illnesses. Sometimes, however, when the *qi* of the body is insufficient, even medication aimed at removing the root cause may not work. This can happen in the case of diarrhea, dysentery and bronchial asthma. Under such circumstances, *qi*-replenishing medicines make up for the deficiencies, thereby improving the overall conditions of the body and consolidating the body resistance. Eventually, the body resistance eliminates the root causes, alleviating the disease or even curing it. Such is known as "using the same treatment for different diseases".

This does not mean, however, that any two similar symptoms can be treated with the same method. Rather, the key is to identify the most important characteristic common to all symptoms and eliminate it. A bad appetite or insomnia, for example, is rarely taken as an essential characteristic. This is because these symptoms rarely, if ever, touch on the nature of the disease. Rather, deficiencies of kidney and *qi* and blood stasis are usually cited as the major characteristics that are the root causes of diseases. It takes experience and expertise to discern those fundamental characteristics in therapy.

# Chapter 41

## Good Timing

Military theories dictate that it is unwise to confront an enemy with high morale and dignified ranks. Even if one is lucky enough to win, victory inevitably has a high cost. Rather, one should wait patiently until the enemy's morale subsides and the ranks collapse. Then use one's elite troops to confront the enemy. In this way, one can win the battle with the minimum cost.

The same principle applies to medicine. Pathogenic factors are the enemy, and body resistance is one's own troops. In therapy, doctors use various means — applying acupuncture, prescribing medicines or performing massages — to reinforce the body resistance and eliminate the pathogenic factors. A good doctor knows the exact timing for the various therapies in order to achieve the best effect. The timing refers to the period when the patient's body resistance is the strongest and pathogenic factors the weakest. Zhang Zhongjing, the father of TCM, used the "ten jujube decoction" to treat excessive body fluids and administered the medicine at the break of day, when the *yang qi* of the body was at its strongest. Modern laboratory tests have also shown that *yang*-strengthening medicines work better in the morning than in the evening. This is because in the morning, the *yang qi* of the body is on the rise.

A good timing is especially important for chronic diseases. When diarrhea first breaks out, for example, pathogenic factors are at their strongest whereas the body resistance is weakest. To subdue the pathogenic factors, it is necessary to have strong medication. Strong medicine, however, also hurts the already weak body resistance mechanism. Therefore, doctors are very cautious in prescribing such medicines.

On the other hand, when diarrhea lies dormant before erupting, pathogenic factors are at their weakest, while the body resistance is the strongest. A small dose of strong medicine or a large dose of mild medicine will do the job — dispelling the pathogenic factors, while preserving the body resistance. Therefore, *The Yellow Emperor's Canon of Internal Medicine* cautions against applying acupuncture or medicine when the pathogenic factors inside the body override the body's resistance mechanism. This principle does not apply, of course, to contingent non-intermittent diseases that have no good "timing" to speak of. Patients with these diseases are advised to seek immediate treatment.

## Chapter 42

## Exploiting the Circumstances

According to the *Records of the Historian*, those who are adroit at war know how to turn the balance of power to their advantage. Likewise, in medicine, the doctor must know how to exploit the balance between body resistance and pathogenic factors inside the body and turn the situation to his advantage. To dispel pathogenic factors, for example, the doctor must base his treatment on the nature, characteristics and location of the factors. The "wind" pathogen, for example, has the characteristics of "lightness" and "dissipation", inclined to invade the upper part of the body and the skin. Correspondingly, in treatment, diaphoretics and drugs capable of inducing sweat and dispelling exopathogens must be used. Common colds due to pathogenic wind-heat are usually marked by fever, headache, sore throat, redness over the tip and margin of the tongue and coughing. In therapy, Yinqiao San, or powder of lonicera and forsythia, must be used to dissipate the wind-heat. Measles, likewise, results from wind-pathogen and can also be cured with wind-dispelling drugs.

By comparison, damp-pathogen is "heavy" and typically invades the lower part of the body. Diseases caused by damp-pathogen should be cured with drugs inducing diuresis and removing dampness.

After invading the human body, pathogenic factors often transfer to different directions — from the exterior to the interior and vice versa; from the top to the bottom and vice versa. Doctors should adjust treatment in light of the direction of such transfers. Indigestion as a pathogen, for example, tends to move upward or downward. A downward movement will result in a swelling abdomen, constipation and bad breath. Under the circumstances, the doctor should follow the downward trend and administer purgative drugs to achieve relief through defecation. Sometimes, though, the pathogen cannot be relieved because the patient has difficulty defecating due to insufficient fluid in the large intestine. In such cases, the doctor should not only administer purgatives, but also fluid-increasing drugs such as ophiopogon root and figwort root.

On the other hand, if the pathogen moves upward, the patient will feel like vomiting. This calls for the use of vomit-inducing drugs such as guadi san, or power of muskmelon pedicel, to dispel the pathogen through vomiting.

The advantage of using the strategy of "exploiting the circumstances" is that it can relieve pathogens in the simplest and fastest way without hurting the body's resistance system. This, of course, is conducive to therapy and recuperation.

## Chapter 43

# Costs and Benefits

In addition to weakening the enemy through armed attacks, another effective way of preserving one's strength in battle is to seek a truce. In TCM, "seeking a truce" is also one of the eight major therapeutic methods (sweating, vomiting, emptying, pacifying, warming, clearing, dispelling and invigorating).

Each of these eight methods applies to different situations. When the pathogen affects primarily the surface of the body, with symptoms such as cold, skin and external diseases, the method of "sweating" (using sweat-inducing drugs) should be adopted; when the pathogen is identified to be in the upper part of the body, such as phlegm in the chest and swallowing toxic substances by mistake, the method of "vomiting" (using vomit-inducing drugs) should be adopted; when the pathogen is in the lower part of the body, the method of "emptying" (using purgatives to empty the intestines) should be adopted; when there is an insufficiency of *yang qi* inside the body, the patient feels cold in the limbs, and defecation is clear, the method of "warming" (using warm medicines) should be applied; when the patient suffers from a heat syndrome with symptoms ranging from a fever to thirst, reddened face and yellowish urine, the method of "clearing" (using cold-natured drugs to clear away the heat pathogen) should be adopted; when the patient is weak and body resistance is ineffectual, the patient will experience fatigue, seminal emission, amenorrhea, dizziness and deafness. Usually, these symptoms are associated with deficiencies of *yin*, *yang*, *qi*, blood or the visceras. Under such circumstances, the method of "invigorating" should be used; when the patient suffers from indigestion with no appetite, a swelling stomach, bad breath and constipation, or when the patient experiences accumulation of swollen masses inside the body due to stagnation of *qi* and blood, the method of "dispelling" should be used. In the case of the former, "dispelling" refers to drugs helping digestion; in the case of the latter, it implies the use of drugs promoting the flow of *qi* and blood to dissipate the swollen masses.

The above seven methods apply to a variety of symptoms involving the exterior, the interior, cold, heat, deficiencies and excesses. However, some diseases lie between the exterior and the interior and between deficiencies and excesses of body resistance. None of the above-mentioned seven methods apply to these diseases, either because the effect of the medicine cannot reach the desired part of the body or because there is no way of dispelling pathogenic factors without damaging the body resistance system.

To find a remedy to these diseases, doctors drew on military strategies and came up with the method of "seeking a truce", whereby the pathogenic factors are dissipated gradually, and body resistance recovers gradually. Eventually, the flow of *qi* and blood is regulated, and the disease is treated. When somebody has had a cold for several days, for example, the initial symptoms of coughing, headache, chill and fever might have already disappeared, but other symptoms remain, such as a bitter taste in the mouth, a dry throat, dizziness, fits and starts of chill and fever, nausea, loss of appetite, feeling of fullness in the chest and dysphoria. This is known in TCM as "*shaoyang* disease", a syndrome with the pathogenic actors located halfway between exterior and interior. The remedy for it is called "xiao chaihu tang" or minor decoction of bupleurum, which is a typical "pacifying" drug.

The real aim of "pacifying", of course, is not to make peace with pathogenic factors, but rather to remove those factors through preservation of body resistance.

# 《孙子兵法》与养生治病

兵法专家、军事博士生导师：吴如嵩
中医专家、医学博士生导师：王洪图
军　医　医　师：黄英

新世界出版社

吴如嵩，贵州铜仁市人，1940年8月生。1962年毕业于贵阳市师范学院中文系。现为中国人民解放军军事科学院战略部中国历代战略研究室研究员、博士生导师。任中国孙子兵法研究会副会长，中国国际文化交流基金会、中国先秦史学会和中国农民战争研究会理事。主要著作有：《孙子兵法新论》、《孙子兵法浅说》、《孙子兵法·孙膑兵法》、《孙子校释》（合作）、《中国古代兵法精粹类编》（主编）、《孙子兵法辞典》（主编）等十余部，还发表了军事学术论文50余篇，获国务院首批颁发的有突出贡献特殊津贴证书。

王洪图，1937年4月出生于天津市蓟县，1963年北京中医学院（现更名为北京中医药大学）医疗专业毕业后留校任教至今。现为该校学术委员会委员、内经教研室主任和博士生导师，兼任中国国家中医药管理局内经重点学科主任和中国中医药学会内经专业委员会主任委员。对中医第一经典《黄帝内经》有深入研究，在理论方面多有创见；临床治病既能遵经守度，又常能"出奇制胜"，尤擅长诊治精神神经性疾病。已发表学术论文40余篇，出版专著13部。1993年获中国国家教育委员会"全国优秀教师"奖章。

黄英，女，1959年9月出生，1977年入伍，现为中国人民解放军某医院军医，从医十五年，有丰富的临床工作经验，先后发表医学论文十余篇。

# 编者的话

吴显林

  距今约 2500 年前中国春秋时代成书的《孙子兵法》，内容精博深邃，被誉为"兵学圣典"。它精辟的世事哲理，深奥的思想方法，已被人们广泛运用在军事以外的企业管理、商业竞争和体育竞赛等诸多领域。但是将兵学运用于医学，还要首推中国中医。中国古代的医学家——战国时的扁鹊、唐代的孙思邈、明代的张景岳和清代的徐大椿等——以兵论医，提出了"防病如防敌"、"治病如治寇"、"用药如用兵"、"下药如用刑"等医理，立论新颖独到，说理精辟透彻，曾造福于人类。

  当代《孙子兵法》专家吴如嵩先生，中医教授、博士导师王洪图先生和医师黄英女士，继承了中国古代医学家的学说，同时发挥各自的学识和专长，发展了前人的医学论断，联手撰写出这部"兵医"同源的书稿，这是中国二千多年来第一部以兵论医的专著。

  本书四十四篇，篇篇均阐述了"兵与医"二者之间的辩证联系。文章短小精悍，内容充实；既各篇独立成章，又相互连成一体；不乏经典病例，又含新人意识；哲理跃然纸上，读来趣味盎然，可谓寓科学性、知识性、趣味性、可读性和实用性于一体的一部精品，宜广大读者珍存，闲时翻阅，调养情志，延年益寿。中西医人士更值得一读，如能在前人的基础上将兵学进一步应用于医学，定能创造出前人未所创造的医学奇迹，这是编著者的期盼，也是人类的期盼。

# 目 录

本书之乡导——浅释阴阳、五行、邪正、虚实 …………… 1
引言 …………………………………………………………… 10
 1、天地之间,莫贵于人 ………………………………… 15
 2、禁祥去疑,信医不信巫 ……………………………… 18
 3、防于未乱,治于未病 ………………………………… 21
 4、兵家重防微,医家重杜渐 …………………………… 26
 5、兵家尚武德,修身重养性 …………………………… 29
 6、地势兵之助,地利人增寿 …………………………… 32
 7、兵非多益,食非过益 ………………………………… 35
 8、养兵莫贵于习练,养生莫善于习动 ………………… 38
 9、一张一弛是文武之道,劳逸适度为养生之宝 ……… 41
 10、怒而兴师必败,愠而处世戕身 ……………………… 44
 11、心静利于制胜,情怡益于永年 ……………………… 48
 12、养兵须作息有序,养生当起居有常 ………………… 52
 13、利胜者辱,过养则病 ………………………………… 56
 14、欲不可禁,欲不可纵 ………………………………… 58
 15、治军要激励士气,养生要勤练气功 ………………… 63
 16、战必选将,疾当择医 ………………………………… 65
 17、官兵一体,医患同心 ………………………………… 68
 18、相敌以达变,四诊以察因 …………………………… 70
 19、运筹以决胜,辨证而施治 …………………………… 75
 20、先伐谋后伐兵,先食疗后药疗 ……………………… 79
 21、胜敌在得法,治病如治寇 …………………………… 82
 22、排兵布阵,用药组方 ………………………………… 86

23、正合奇胜巧应变,君臣佐使须分明 …………………… 89
24、用兵宜慎,服药戒躁 …………………………………… 93
25、攻敌先攻心,治病先治心 ……………………………… 97
26、抑情制怒,以情制情 …………………………………… 100
27、乖其所之,移情易性 …………………………………… 103
28、兵不厌诈,疑病疑治 …………………………………… 107
29、气失则师散,神衰则体病 ……………………………… 110
30、用乡导得地利,依归经利药效 ………………………… 113
31、擒贼擒王,治病治根 …………………………………… 117
32、老敌之师,扶正祛邪 …………………………………… 120
33、断敌要道,阻隔病源 …………………………………… 123
34、守我岩疆,先安未病 …………………………………… 125
35、焚敌资粮,断除病源 …………………………………… 128
36、断敌内应,标本兼治 …………………………………… 131
37、地有六形,医有六经 …………………………………… 133
38、治齿如治军,治目如治民 ……………………………… 135
39、以寡胜众,同病异治 …………………………………… 137
40、捣敌中坚,异病同治 …………………………………… 140
41、无击堂堂之阵,无刺熇熇之热 ………………………… 143
42、推波助澜,因势利导 …………………………………… 145
43、权衡得失,和解为宜 …………………………………… 147

# 本书之乡导——浅释阴阳、五行、邪正、虚实

中国医学理论体系的形成与《孙子兵法》的诞生,处在相同的历史时代,两者受着同一时代哲学思想的影响,带有相同的时代烙印。因此,虽然一者论医,一者论兵,但在认识问题的方法甚至讨论问题所用的术语方面都有许多相同或相近之处。如阴与阳、虚与实、攻伐与调养、邪气与正气等等,本书所辑43篇,全部可以反映出医与兵二者之间的密切联系。

然而,由于时代的变迁,语言也发生了重大的变化,今天的人们与古人在世界观和方法论方面已有很大不同,要想学通古汉语,还非得下一翻苦功夫不可呢。正因为这样,本书所引用的古代兵书和中医学的某些特定词语,尤其是专业术语,对大多数读者而言,很可能是颇为生僻的。故特写此浅释阴阳、五行、邪正、虚实篇,以为阅读本书之向导。

## 浅释阴阳

在春秋战国时代,盛行着一种哲学,它用阴和阳来认识世界,用阴和阳做为分析和解决具体问题的方法,后来学术界称之为"阴阳学说"。《黄帝内经·素问·阴阳应象大论》对阴阳的意义做过如下含意的说明:"阴阳是宇宙间的普遍规律,是分析和归纳万千客观事物的纲领,是一切变化的根源,也是发生、发展、消亡的内在根据。所以说,尽管世界有无穷无尽的奥妙,但都是从阴阳演化出来的。因此,诊断和治疗疾病,就必须从阴阳这个根本问题上着眼。"

1

进而言之,包括行军安营、排兵布阵在内的一切行为,当然也必须根据阴阳行事,才能立于不败之地。

**阴阳学说包含有以下几个基本观点:**

**第一是对立、制约和消长**。阴与阳作为一对哲学范畴,可以把无限的具体事物提纲挈领地划分为阴与阳两个方面;同一事物内部也含有阴与阳两个方面。一般地说,凡具有光明、上升、活跃、充实、在外、热、轻清等性质和特点的事物,都归属于阳。相对而言,凡具有黑暗、向下、安静、柔弱、在内、寒、重浊等性质和特点的事物,都归属于阴。如天在上而轻清为阳,地在下而重浊为阴。日光明常满而热称为太阳,月光寒而常缺称为太阴。火性热而上炎为阳,水性寒而下流为阴。男子刚勇,精气应充实为阳,女子温柔,月经依时而下为阴……在同一人体中,心与肺在上部为阳,肝、脾、肾在下部为阴。人身之气无形而活跃为阳,人身之血有质、流行于脉中为阴。人身有十二条主要经脉,六条行于肢体外侧为阳经,六条行于肢体内侧为阴经。每一个脏俯,其形态部分可归属于阳,其功能活动部分可归属于阴,如心阴、心阳、肾阴、肾阳、脾阴、脾阳等。

由于阴与阳具有相互对应的性质,因而必然相互制约、相互消长。如寒多热必少,寒少热必多。一年中春、夏季自温而热为阳,寒气渐少;秋、冬季自凉而寒为阴,热气渐衰。此方长则彼方消,此方消则彼方长。例如高烧的病人是"阳长",常常出现口干、皮肤起皱等"阴消"的现象,治疗除应清其阳热外,还应注意补充水分。有时病人虽然也发烧,表现为"阳长",但却是因为"阴消"所造成的,由于阴发不足而不能制约阳长所致。治疗就应以补阴为主,使水能制火(阴能制阳),其热自退。

**第二是统一、互根**。与对立不可分离的是统一,中医学里叫"阴阳互根"。互根有两重意思,一是互为根据、互为存在的前提,失去一方,另一方便不复存在,或失去原有的意义。如无寒便无所谓热,无外也无所谓内,没有失败就无所谓胜利。人体如果没有气(阳)形体(阴)也就失去意义,当然没有形体则气也必定荡然无存。因此,健康就是形与气都旺盛,阴与阳相互协调的状态。治病的道理,从根本上说就是调和人体内的阴阳。互根的第二层意思是阴阳相互转化。"重阴必阳,重阳必阴","否极泰来"都是讲的转化。如春、夏由温到热(属阳),但到夏至阳气最盛(地球距太阳最近),物极必反,便开始转化,阴气渐生。到冬至(地球离太阳最远),阴气极盛,转化为阳气渐生。所以古人说"冬至一阳生","夏至一阴生"。转化是需要条件的,比如以较弱小的军队战胜较强大的敌人,使敌我双方在势力上发生转化,其条件必须是我方指挥得当,或敌方明显失误。有些高烧、面红、脉象有力的病人(阳盛病,又叫阳症),如果治疗不得法,也可能会突然体温低于正常、面色苍白,脉象无力,转化为阴证。

**第三是阴阳中复有阴阳、阴阳交合**。事物的阴阳性质是相对的,如昼为阳,夜为阴,但上午为阳中之阳,下午为阳中之阴;前半夜是阴中之阴,后半夜是阴中之阳。女人中也有"强人",男子中也不乏弱者。天为阳,但天也有下降之势,才能生成雨露。地为阴,地气也必须上升,才能有云彩。天地阴阳相互交合,才能化生万物。《黄帝内经·素问·天元纪大论》说:"天有阴阳,地亦有阴阳……动静相合,上下相临,阴阳相错,而变化由生。"《易传·系辞》说:"男女构精,万物化生",都是讲阴阳中复有阴阳,阴阳相交合化生万物的道理。

## 浅论五行

五行,简言之就是木、火、土、金、水的运行,它是中国古人从实践中抽象出来的哲学概念,用来说明世间一切事物之间的关系,后来学术界称之为"五行学说"。五行的原始含义,是客观存在的五种基本物质,即人们饮食所用的水与火、生长植物的土地、经冶炼可以做成各种器物的金属、蓬勃生长的草木。对这五种物质的特性加以再认识,抽象出一般的概念,用以理解世界万物的起源和多样性的统一,形成为一种哲学思想,而盛行于春秋战国时期。在五行学说中,各行的基本特性如下:

**木**:柔韧向上伸展,像春天和风吹拂的青草树木,生命力看似柔弱却是任何力量也不能扼止的。比如人体中的肝脏,含有极强的活力,能使人心情舒展。又如人体的筋,柔韧而有力量。

**火**:炎热活跃而向上燃烧,像夏季火红的太阳使万物生长茂盛。比如人体中的心脏,推动血液不停地流动,营养着全身;心脏所藏的精神活跃而不易平静。又如人的舌灵活多动。

**土**:墩厚宁静,既能容纳万物,又能滋养万物,而为万物之母,像夏末(长夏)时植物繁茂,孕育果实。比如人体中的脾脏,吸收饮食物中的营养,生成气血津液,供给生命活动所需要的物质。脾脏功能旺盛,肌肉才能丰厚。

**金**:清凉而能制成各种利器,具有一种肃杀之气,像秋天的凉风,使草木凋谢而下落。比如人体中的肺,主持呼吸,其气必须下降,否则气上逆就引起咳喘。

**水**:沉静寒冷,有向下的趋势,像冬天气候寒冽,蛰虫潜藏。比如人体中的肾脏,藏有生殖之精,此精也应蛰藏而不宜妄泻。肾精充足,骨质才能微密。老人肾虚,骨质必然疏松。

**五行之间的关系有两方面:**

相互滋生助长,叫"相生"。即木生火,火生土,土生金,金生水,水生木,周而复始。

相互制约,叫"相克"。即木克土,土克水,水克火,火克金,金克木,循环往复。

本书第34篇引用古人见肝之病则知当传之于脾之语,这个认识除实践经验之外,还基于"木克土"的哲学观点。因为肝属木,脾属土,木能克土,所以才"见肝之病,知肝传脾",而及时加以防范。

五行学说做为古代哲学,它力图把宇宙间一切事物都加以概括,如:

| 五行 | 方位 | 季 | 五气 | 色 | 生化 | 味 | 星 | 脏 | 音 | 声 | 体 | 畜 | 干 | 支、…… |
|---|---|---|---|---|---|---|---|---|---|---|---|---|---|---|
| 木 | 东 | 春 | 风 | 青 | 生 | 酸 | 木 | 肝 | 角 | 呼 | 筋 | 鸡 | 甲乙 | 寅卯…… |
| 火 | 南 | 夏 | 火 | 赤 | 长 | 苦 | 火 | 心 | 徵 | 笑 | 脉 | 羊 | 丙丁 | 巳午…… |
| 土 | 中 | 长夏 | 湿 | 黄 | 化 | 甘 | 土 | 脾 | 宫 | 歌 | 肉 | 牛 | 戊己 | 辰戌丑未… |
| 金 | 西 | 秋 | 燥 | 白 | 收 | 辛 | 金 | 肺 | 商 | 哭 | 皮 | 马 | 庚辛 | 申酉…… |
| 水 | 北 | 冬 | 寒 | 黑 | 藏 | 咸 | 水 | 肾 | 羽 | 呻 | 骨 | 猪 | 壬癸 | 亥子…… |

各行之间的事物,是彼此联系而存在的,离开联系,便无意义。如东方,是与其它四方相联系而存在。孤立地看待某一事物,便无"五行"可言。

由于阴阳学说是以论"一分为二"对立统一规律为主,而五行学说是以论事物相互联系的规律见长,故两个学说在发展过程中,逐渐结合,互为补充,而被称为"阴阳五行学说"。在古代,这个学说被应用到天文、地理、历法、美术、农业、水利、军事、医学各个学科。时至今日,在中医学里仍然较完整的保留和应用着,这是因为,不仅阴阳五行的哲学观点有其可取之处,更因为它早已和医学的具体内容结合在一起,而并非仅仅是哲学问题。如前所说肾阴、肾阳、心阴、心阳、阴经、阳经,以及肝病传脾用木克土代之等,很难从中医学中剔除。从事西医工作的中国医生,也难免受到中国传统文化的薰陶,常把化验、检查结果的"有"称为阳性、"无"称为阴性。

5

# 略论邪正

邪与正是相互对等之词。正，又叫正气，是指善良、正义、美好以及一切有益于社会、有益于人体健康的事物。如军事上为反侵略、反压迫而战的叫正，或正义战争。在中医学里，正气分自然界和人体内部两种：风、寒、暑、湿、燥、火六种气候，在正常的变化范围之内，是包括人类在内的一切生物赖以生存的条件，叫做"六气"，便属于正气；人体内的气、血、津、液、精、神，是生命活力的正常物质和功能，也是抵御不利因素侵扰的力量，所以统称为人体的正气。

邪，又称邪气，与正气相反，是指险恶、凶残、非正义以及一切有害于社会、有害于人体健康的事物。如发动侵略战争的势力、犯毒行为、危害人们身心健康的物质和精神产品、引起人体生病的一切因素等。在中医学里，邪气也分为自外界侵犯人体的和体内自生的两类。上述风、寒、暑、湿、燥、火六气，如果变化失常，超过人体所能适应的限度而引起发病，便是邪气，而称为"六淫"，或叫外感邪气。饮酒过量、食物不能消化或误食毒物，也危害健康，引起疾病，这些饮食、毒物也属于邪气之类。人体自身也会产生邪气，如因劳累过度、精神紧张等，使脏腑功能（气）失调，产生心烦、失眠、牙龈肿痛、便秘等症状，中医统称为"上火"，这个火也是一种邪气。火从何来呢？其实就是身体某些功能（气）过于亢奋，从正气转为邪气了，医学上把这种情况叫做"气有余便是火"。此外，当人体有某些疾病时，由于生理功能紊乱，也会相继产生痰饮、瘀血、水湿等，这些虽然都是病理产物，但反过来又能造成新的疾病，所以痰饮等也属于"邪气"。

不难看出，邪与正是相对的，本是同一事物，当其在一定范围内时，可以是正，但超过限度就成为邪气。如血是生命物质之一，如果瘀滞了也就成为邪气。从医学角度看，相对性还表现在人们的个体条件有关，例如同一气候变化，张先生能适应而不生病，李先生不能适应便生了病，那么这个气候是正是邪呢？对张而言是正，对李而言则是邪。至于张不病而李病的原因，主要是两人体内的正气有旺盛与虚弱之分，《黄帝内经》所述"正气存内，邪不可干，"明确地回答了这个问题。因此，养生防病的关键，就在于通过多种方法来培养人体的正气。但应切记，事物还有另外一面，即尽管正气充足，身体健康，假若遇到特殊而剧烈的邪气，恐怕也难幸免于病，所以《黄帝内经》又告诫人们说："虚邪贼风，避之有时。"举例言之，与其挖空心思去寻求解酒药，不如少饮为佳。既然惧怕艾滋病，就不要涉足是非之地。

## 简析虚实

虚与实也是相互对立之词。实，指有余、充足、旺盛；虚，指不足、衰弱、短少。人们对将要发生的事，做了充分准备，成竹在胸，就会心里"踏实"；而对无准备、无把握的事，难免心中"空虚"。军队作战常采用"避实就虚"的方法，为了迷惑敌人也常用"实者虚之，虚者实之"的战术。蜀汉将领张飞在当阳桥的"虚张声势"，退去数十万曹兵；诸葛亮用"虚者实之"空城计，挽救危亡。

在中医学里，除从总体上判断人的健康状态时，分为壮实与虚弱之外，就诊断疾病而言，也把所有疾病分为虚、实两类，称为虚证、实证。所谓实证，其基本特点是邪气过盛。不论外感邪气，还是内生的瘀血、痰、火，都属于此类。治疗实证的总原则是用泻法，叫做"实则泻之"。

如外感病，实邪在皮表，用发汗药治疗；毒物或不化的食物在胃中，用催吐药；燥屎在肠中，大便不通，用攻下药;瘀血停留,用活血祛瘀药;火热邪气在体内,用清热泻火药;水邪停留体内,浮肿、小便不利,用利尿药等,都是因为有实邪而采用的"泻"法。

五脏病也有实证,如心火旺,出现舌上生疮、尿赤、尿痛,用导赤丸治疗以泻心火;肝脏有实热,出现急燥易怒、头晕、耳鸣、大便干燥,用泻青丸治疗;肾脏有热,出现失眠、遗精、性欲亢进,用知柏地黄丸治疗,都属于"实则泻之"之列。

虚证的基本特点是人体的正气不足,或称正气虚。和实证一样,虚证也应分清人体何处之虚,在"虚则补之"的治疗原则指导下,分别使用补药。

如疲倦无力、语声低微、面色苍白、脉象软弱无力是气虚证,可以用参苓白术丸治疗;面色萎黄、心慌、失眠、月经量少、脉象细软是血虚证,用四物汤治疗,阿胶浆是由驴皮熬制而成的,也有补血的作用;手足心热、两颧潮红、盗汗、脉象细数是阴虚证,可以用六味地黄丸治疗;四肢清冷、大便稀溏、小便清长、脉象沉迟是阳虚证,可以用金匮肾气丸治疗。

五脏病,也各有虚证,医生根据病人的临床症状,既要判断病在哪一脏,还要判断有病之脏是气虚、血虚,还是阴虚、阳虚。如心脏虚证,分别为心气虚、心血虚、心阴虚、心阳虚等证,而给予不同的补药来治疗。

虚证与实证虽然种类繁多,但对有相当水平的中医来说,并不难辨认。可是还有一些病人,表现出既有正气虚一面,又有邪气盛的一面,成为"虚实挟杂证",在分辨正与邪孰多孰少的比例上,较费周折。虽然治疗这类病证的总原则是"攻补兼施",但究竟应该用多少补药、用多少泻药？或者是应先补后泻、还是应先泻后补？要把握好分寸,确实要求医生有较高的水平。以癌症为例,那个令人可怕、可厌的肿瘤,当属实邪无疑,应该用泻药去削除它。但病人的正气也已多半虚弱,又应该用补药去救助它。可是,治实邪的泻药多半都有损伤正气的副作用,而补正气的药物又可能有助长病邪的副作用。要能权衡出利弊、分清其主次来,实非易事啊！

从诊断疾病的角度而言,还有比虚实挟杂证更难解的题目,那就是"假像"。疾病本是实证,但却表现出若干"虚"的现象;或者原是虚证,反而表现出不少"实"的症状,这更要检验医生的水平了。古人曾提醒医生说:"至虚有盛候,大实有羸状",千万注意,不要被假像所迷惑。例如妇女有瘀血,却表现出月经不来、身体羸瘦、皮肤干燥等"虚"的现象,如果误认为虚证,那就大错而特错了。其实,应该用大黄䗪虫丸破其瘀血才能见效。当然"假像"的背后,还是有真像可以察到的,如上述瘀血的女病人,她的舌上必定有紫色斑,或者全舌发紫,她的脉象必定坚实而不虚软,便是真像。尽管医学书籍中反复强调,但在临床上出现误诊的事例仍时有发生。以曹操之精明、司马懿之狡诈,不是也吃过"假像"的亏吗？当然,主要还是由于他们疏忽大意所致。

# 引 言

公元 18 世纪中叶的清朝乾隆年间,有一位名医徐大椿。他原名大业,字灵胎,江苏吴江人,曾任太医院的太医。一生著述宏丰,著有《经释》、《类方》、《慎疾刍言》等,后人评论其书"推阐主治之义,于诸家中最有启发之功。"值得一提的是他的论医之书《医学源流论》,其中辟有专章《用药如用兵论》。他在全面、准确地阐述用药如用兵的医理之后,明确地指出:"《孙武子》十三篇,治病之法尽之矣。"

徐大椿凭藉自己深厚的医学功底,敏锐地审视兵学与医学、兵道与医道的相互关系,得出了极富启发意义的结论,这在中国古代医学史上具有不可低估的学术价值。但是,必须看到,把兵学同医学联系起来在理论上加以阐述,临床上进行实践,则在徐大椿之前是大有人在的。

被尊为医经的《黄帝内经·灵枢·逆顺》中就谈到医学与兵法的关系,说:"无迎逢逢之气,无击堂堂之阵。"以与"无刺熇熇之热,无刺漉漉之汗,无刺浑浑之脉"相对应。饶有意味的是,《黄帝内经》还用冷兵器的"五兵"(弓矢、戟、矛、戈、殳)类比针法:"两军相当,旗帜相望,白刃阵于中野者,此非一日之谋也;能使民令行禁止,士卒无白刃之难者,非一日之教也,须臾之得也"(《灵枢·玉版》)。

唐朝杰出的医学家孙思邈进一步从医家的行为心理上做了深刻的理论阐述:"胆欲大而心欲小,智欲圆而行欲方"(《旧唐书·孙思邈传》)。"胆大心小,智圆行方"这八个字高度概括了作为一个医生也必须具备的心理素质和行为准则。孙思邈在对这八字原则进行分析时指出,医生临病与军人临战一样,在不明敌情时要周密侦察,慎重判断,做到"知彼知己",这种谨慎就是"心小"。一旦掌握敌情,胜券在握,就要果断决策,大胆用兵,这种果断就是"胆大"。孙思邈用《诗经》形容这种果断:"'纠纠武夫,公侯干城',谓大胆也。"军人为卫国保民而打仗,医生为救死扶伤而治病,这种品德就是"行方"。用兵要因敌制胜,战术多变;用药要"知常知变,能神能明,如是者谓之智圆"(李中梓《医宗必读》)。

以兵学喻医学,在中国古代并不是个别例子,自先秦至明清,代不乏人,可以说它是古代医家们的一种共识。那么人们要问:兵学原则在哪些方面成为沟通医学的桥梁和纽带呢?毋庸置疑,只有对这个问题作出正确的回答,才能证明兵学用于医学不是牵强附会。

首先,从伦理道德方面看,中医的医学伦理观一贯推崇"良医"(既有良好的医德,又有良好的医术),认为医学是仁学,医术是仁术。东汉名医张仲景主张医生应"上以疗君亲之疾,下以救贫贱之厄"。这种中国医学史上倡导的"苍生大医"与中国古代兵家的主张是相通的,相融的。与行医要有仁人之心一样,中国兵家在对待战争的态度上也历来主张应"诛暴乱,禁不义"(《尉缭子·武议》)。因此,《孙子兵法》要求将帅必须做到"进不求名,退不避罪,唯人是保,而利合于主","仁德"便成为将帅修养的核心。

其次,从方法论上看,中国医学与中国兵学一样,形成了完备而严密的理论体系,都贯穿着朴素的唯物论和辩证法。具体地说,表现为"天人合一"的中医学理论。例如,春秋时良医医和指出:天有"六气","阴、阳、风、雨、晦、明也。分为四时,序为五节,过则为灾,阴淫寒疾,阳淫热疾,风淫末疾,雨淫腹疾,晦淫惑疾,明淫心疾"(《左传·昭公元年》)。意思是说,阴、晴、风、雨、夜、昼,分为四段时间,五种音调。过头了,不协调就会招灾惹祸。阴过头了是寒病,阳过头了是热病,风过头了是手脚病,雨过头了是腹病,夜里没有节制是迷乱病,白天没有节制是心病。辩证施治是中医著名的诊断治疗方法。从哲学上说,它是一种朴素的系统的思维方法。中医把天地人视为一个大系统,"人"又分为五脏六腑、十二经脉各个子系统。通过望、闻、问、切,对各个系统进行综合分析之后,对症下药。药分君臣佐使,最后达到治病求本,标本兼治,扶正祛邪,治病救人的目的。

　　古代兵法也是使用这种系统分析的方法。比如说,关于战争全局的战略问题,《孙子兵法》讲求"道、天、地、将、法"之类"五事七计",对战争各方面进行总体把握,系统分析;关于行军布阵之类战术问题,兵家也无不讲求天、地、人的相互关联,各种条件在战争运动过程中的有序变化,奇正相生,相反相成,最后达到因敌制胜的目的。

　　应当看到,朴素的系统论不仅是中医学、传统兵学的精髓,也是整个中国学术思想的精髓,它是迥别于西方哲学的。瑞典系统论专家普里高津曾经指出:"中国的传统学术思想是着重于研究整体性和自发性,研究协调和协合。"他的看法是非常中肯的。

　　再次,从指导思想上看,医学与兵学更是有着许多共同点。
　　一是防病如防敌。对于疾病,医家主张"圣人不治已病治未病","良医者,常治无病之病,故无病"。只有那些能预防或减少疾病发生的医生,才能称得上是良医。对于敌人,兵家主张,"为之于未有,治之于未乱","天下虽安,忘战必危"(《司马法·仁本》),"故用兵之法,无恃其不来,恃吾有以待也;无恃其不攻,恃吾有所不可攻也"(《孙子兵法·九变》),平时就要有备无患,"立于不败之地"。这样的将帅也才能被称为良将。良医本着治病如治寇的负责精神,因而能够高度重视对疾病的预防,做到防微杜渐,加强防范。

二是择医如用将。明朝褚澄说得好:"知其才智,以军付之,用将之道也;知其方技,以生付之,用医之道也"(《褚氏遗书》)。打仗要委派良将指挥,治病要选求良医诊治,道理是相通的。中国古代所谓良将必须"智信仁勇严"五德兼备,也就是说,必须德才兼备,智勇双全。只有将才而无武德的将领只能称为"名将"而不能称为"良将"。同样,只有医术而乏医德的医生只能称为"名医"而不能称为"良医"。"良医"必须做到"性存温雅,志必谦恭,动须礼节,举乃和柔,无自妄尊,不可矫饰。广收方论,博通义理,明运气,晓阴阳,善诊切,精察视,辨真伪,分寒热,审标本,识轻重"(《小儿卫生总微论方》)。如能选择这样优秀的医生治病,何病不克,何疾不愈?!

三是用药如用兵。兵凶战危,是一种关乎国家军民生死存亡的暴力行为。古代的医家看到,"药性刚烈,犹若御兵。兵之猛暴,岂容妄发"(《千金要方·食治》)! 这是从"兵"与"药"的特性上说明二者都具有"刚烈"的共同特点,因此用药要慎之又慎。

古人还从用药之法"贵乎明变"着眼,看到灵活多变的共同点。徐春甫《古今医统》指出:"治病犹对垒。攻守奇正,量敌而应者,将之良;针灸用药因病而施治者,医之良也。"这是医家以用兵来比喻用药。《白豪子兵畾》指出:“良将用兵,若良医疗病,病万变药亦万变。”这是兵家以用药来比喻用兵。显而易见,“兵”与“医”即使不象徐大椿所认为的同源,但也的确是相通相用的。

以上择其要者,略举数端,旨在说明把兵学原理移植到医学之中,特别是中医学之中,无论在理论上还是实践上都可以启发人们的意智,开阔人们的视野。宋朝哲学家程颐说得好:“天下之理一也,途虽殊而其归则同,虑虽百而其致则一。虽物有万殊,事有万变,统之以一,则无能违也”(《伊川易传》卷三)。他说的用来统率万事万物的“一”,就是哲学。具体地说,就是思维方法。掌握了这种思维方法,就能架起沟通兵学与医学的桥梁。这座桥梁,在中国古人的辞典中称为“悟”。对此,明朝著名学者黄宗羲曾说过一段非常透辟的话:“为学为教,舍自得别无它路。欲自得,舍悟别无它路”(《明儒学案》卷二十六)。钻研任何一门学问,关键要自己确有心得,确实掌握要领,得其精要。之所以能够“自得”,关键又在于“悟”。所谓悟,就在于通过正确的学习,切实的深造,在具有丰富经验和深厚学术功底的基础上,触类旁通,默识心悟,开启新的学术天地。不难设想,倘有精通医学与兵学者,在这个领域深入开掘,未尝不可以创造医学理论的新思维。

## 1、天地之间,莫贵于人

《孙膑兵法·月战》一文写道:"天地之间,莫贵于人。""不得不战才去作战。"讲的是中国古代军事家珍惜人的生命。《黄帝内经·素向》一篇讲道:"天复地载,万物悉备,莫贵于人。"人应该"尽终其天年,度百岁乃去。"讲的是古代医家对人类生命的珍爱。为了人类健康长命,还是在蛮荒时代,中国就有中医始祖神农尝百草、救人生命的动人故事。

据说,当时自然灾害泛滥,人们"一日而遇七十毒",疾病遍地流传,致使民不聊生。有位神农氏,"尝百草之滋味,水泉之甘苦",探究保护生命的方法和规律。后来,轩辕氏(黄帝),在发明文字、舟车、音律、算数、养蚕的同时,对医学做出了重要的贡献。黄帝有两个医臣,一个是岐伯,一个是雷公。岐伯善于经方治病,雷公长于针灸祛疾。他们君臣三人经常在一起深入研究人与天地万物的关系,考察五气变化的规律,分析阴阳转化的特点,洞悉人类生老病死的过程,探寻治病祛疾的药物和方法,然后以君臣问答的方式写出医学巨著——这就是中外闻名的典籍《黄帝内经》。

中国医学的始祖自然当首推神农、轩辕等。中医学自从它产生之日起,就融合了贵人贱物的思想传统,是以治病救人为宗旨的仁术。唐代医学家孙思邈在《千金要方》一书的序言中说,"人命至重,有贵千金,一方济之,德逾于此。"为了达到治病救人的目的,他又从医德的角度提出了三条具体原则:

一、治病必须具备仁爱之心,淡泊名利。凡行医者必须象古代良医那样，潜心研究医术，消除任何物质欲望和名利奢求，怀藏大慈恻隐之心，以救死扶伤为神圣职责，甚至不惜以自己的生命去解救病人的危急。决不能凭着一技之长，醉心于索取钱财。

二、治病不分亲疏贵贱，一视同仁。如果有患者前来求治，不论贵贱贫富、长幼妍媸、恩怨亲疏、华夷愚智，都应像亲人一样热情对待，诚心救治。决不能嫌贫爱富，偏心偏向。

三、治病应不辞艰辛，急病人之所急。救死扶伤过程中，不能瞻前顾后，过多考虑自己的名利得失。应当将患者的疾苦当作自己的疾苦，不论道路险恶，不分昼夜寒暑，不管饥渴疲劳，一心赴救，决不能拖延慢待，敷衍了事。

孙思邈认为，一个医师只有从这三个方面严于律己，才可能成为真正以人为贵的苍生大医。反之，则是人类的巨贼。

这些原则既是对前人思想的概括，又是对历代良医经验的总结。中国历史上著名的良医莫不如此，流传至今的"杏林春暖"的故事就颇有代表性：

三国时期，江西庐山隐居着一位名叫董奉的医生。他对求治的患者，不论疾病轻重，从不拒绝，并且在医治过程中不取分文，但对病愈后前来酬谢的人却有一个特殊的要求。那就是根据来者原来病情的轻重，让他们分别在自己的房前屋后种植数目不等的杏树。病轻的人种一株,病重的人种五株。数年之后,董奉隐居之处竟然杏树成林,郁郁葱葱。早春时节,杏花盛开,春色满园。初夏之际,杏黄大熟,硕果累累,清香扑鼻。每当杏熟时,他在林中的谷仓里设一器具,张榜通告过往行人,有欲买杏者,可按规定,一器谷换一器杏,自行取去,不必通报。对于所换得的大量粮食,董奉不是用来发财致富,而是除留下自己的口粮外,绝大部分用于接济贫困孤寡和无依无靠的老人,以及行旅不逮之人。

董奉这种以人为贵、治病救人的生动事迹,为人们交口称道。随着董奉美名的传扬,杏林的佳话不胫而走,遍及天下。"杏林春暖"、"誉满杏林"、"杏林春满"、"杏林望重"等等,成为古今病家借以表达对医家感激和尊敬之情的共同语言。古今许多医家则以"杏林"自励,以致出现过一些以杏林命名的医院、药房和医学团体,诸如杏林医院、杏林堂、杏林学社等等。

从"尝百草"到"杏林春暖",再到当代盛赞的"白求恩精神",这其中虽然间隔着千百万年,但是"天地之间,莫贵于人"的思想是一脉相承的。这是中国医学始终一贯的宗旨,也是中国历代良医奉行不悖的最高道德准则。

## 2、禁祥去疑，信医不信巫

远古时代，慑于自然力的威胁，先民们的认识世界里被神秘恐惧笼罩着，几乎是一个巫术的世界。无论是战争，是人事，是生活，人们的思维和行为无不蒙上巫术的阴影。

但是，在军事领域里，荒诞不经的符咒逐渐被人们识破。兵学家孙武从无数历史事实中清楚地看到，战争是力量的拼搏，智慧的较量，最终的胜负取决于人的智慧和力量，而不是取决于神的护祐，提出了"禁祥去疑"的主张。"祥"是古代祸福吉凶预兆的通称。"禁祥"，即不得以占卜问筮等各种迷信活动疑惑人心。而且，他还要求将帅侦察敌情时，"不可取于鬼神，不可象于事，不可验于度，必取于人"，取于那些了解敌情的人，因为战争的胜负关系着国家的安危和军民的生死。

同样，医学关系到人们的健康与生死，在医学产生与发展之初，排除迷信观念显得尤为重要。成书于两千多年前的中国第一部医学典籍《黄帝内经》这样写道："拘于鬼神者，不可与言至德。恶于针石者，不可与言至巧。"至德，就是指医学理论；至巧，是言针刺、砭石等治疗疾病的技术与措施。"不可与言"，表明医学科学与迷信鬼神的观念不可同日而语，誓不两立。

战国时名医扁鹊，勃海郡郑人，姓秦，名越人，他医术高明，专心致力于为群众治疗疾病，从不计较名利地位，声名远播。汉代历史学家司马迁在撰写《史记》时，特为他立传。因此，扁鹊是我国正史中第一位有传的医学家。他曾提出"六不治"的原则，其中明确宣称："信巫不信医不治"，这对当时的有神论是一个有力的挑战。

《史记·扁鹊·仓公》中记载着一则颇为有趣的故事。一次扁鹊路过虢国，虢国太子病死已有半日，大臣们别无他法，只是祈祷，请求神仙让太子转世回生。扁鹊闻说此事，即往王宫探视。在问明发病情况后，扁鹊说他能让太子起死回生。一个名叫中庶子的大臣对扁鹊说，"我听说古时为医的俞跗，能按治脑髓，持取膏肓，令死人复为生人，你难道有他那样的技法吗？"扁鹊回答说："……子以吾为不诚，试入诊太子，当闻其耳鸣而鼻张，循其两股以至于阴，当尚温也。"中庶子闻言入禀国王。虢君大惊出见扁鹊，说："吾儿有先生则活，无先生则弃捐填沟壑，长终而不得反。"扁鹊告诉虢君，太子患的是"尸蹶"，并没有真死。虢国君主抱着一线希望允许扁鹊医治。扁鹊便立即着手救治。他先命一弟子给太子针灸三阳五会穴，过了一会儿，太子便慢慢苏醒过来。他又命另一弟子用药物热敷，熨贴病人两侧胸胁部，太子终于慢慢坐起身来。接着，扁鹊根据太子的康复情况，每天让弟子给太子按摩、煨汤药。二十多天后，太子终于恢复了健康。众人纷纷称赞扁鹊有起死回生之术。扁鹊却实事求是地告诉大家："越人（扁鹊之名）非能生死人也，此自当生者，越人能使之起耳。"

汉代末年的名医张仲景进一步发展了扁鹊的思想，他在《伤寒论》中说：当今世界上的许多人，不大留心医药，不去深入研究医术，以便上疗君亲之疾，下救贫贱之厄，中保身体健康，而是竞相追逐荣华和权势，企求成为大权在握的豪门贵族，因而孜孜汲汲，唯名利是务。如此弃本求末，欲华其表而悴其内，皮之不存，毛将焉附？其结果必然是既不能爱人知物，也不能珍惜自己。一旦遇上风邪之气、非常之疾便无法抵挡，当病患在身时方才震惊。然而由于平时对医学无知无识，此时便一心企求巫祝祈祷。将百年之寿命，宝贵的生命，委付于巫医，任其摆布和玩弄，自己却只能暗自悲叹。至于身死之后，变为异物，幽潜重泉，哪里还谈得上什么荣华

19

富贵呢?

　　这些饱含唯物思想的言论,既针砭了重名利、轻养生、弃本求末的时弊,又揭示了信巫不信医的恶果。不仅在当时,就是在今天,这些观点也仍然具有震聋发聩的作用。

　　我们平时即使没有患病,也必须学习和掌握一些医学基本常识,以便有意识地调养身心健康,一旦有疾也便于主动与医生配合,增强医治效果。

## 3、防于未乱,治于未病

两千多年前,中国古代的医学家就告诫人们,预防疾病如同预防战争一样,要居安思危。成书于西汉时期的《黄帝内经》明确写道:"圣人不治已病治未病,不治已乱治未乱;夫病已成而后药之,乱已成而后治之,譬犹渴而穿井,斗而铸兵,不亦晚乎!"它生动地把战备国防与疾病预防联系起来加以论述,清楚地说明了防重于治的重要性和以预防为主的保健思想。

中国是一个文明古国,有着悠久的战争历史。成书于三千多年前的《易经》就曾写道:"安而不忘危,存而不忘亡,治而不忘乱。"这是若干历史经验的总结。当时有一个卫公好鹤的故事就是说的忘战亡国这件事。

卫懿公是卫国的国君,平时喜欢养鹤,让鹤享有官位爵禄,乘华车,吃玉食,对国防战备的事却不放在心上。不久,狄国进攻卫国,卫国官兵不愿为昏君卫公卖命,卫国便被灭亡了。

由于历史上这种例子屡见不鲜,所以孙子明确指出:"无恃其不来,恃吾有以待也;无恃其不攻,恃吾有所不可攻也。"告诫人们,不要寄希望于敌人不来,而要依靠自己做好充分准备;不要寄希望于敌人不发动攻击,而要依靠自己已具备强大实力,使敌人不敢发动进攻。

好鹤亡国与小病丧生虽不是同一范畴的问题,但其中的教训无疑是相通的。任何疾病的发生,都有一个从无到有,从小到大,从轻到重的渐进过程。因此,高度重视预防,重视防微杜渐,重视消除病患于未萌或初萌阶段,就具有十分重要的战略意义。

防病真如防敌。防敌要依靠实力强大,防病要依靠身心健康;防敌要居安思危,防病要惜身养生。慎终如始,坚持不懈。有一本叫《嵩山太无先生气经》的书讲过一段话,至今读来仍觉得清新可喜,哲理隽永。他从"惜未危之命,惧未祸之祸,治未病之病"的观点出发,极力倡导"爱精重气",增强体质,搞好预防。

预防战争强调知彼知己,预防疾病也讲究预知致病原因。导致人体产生疾病的原因是多种多样的,中医认为主要有三大类:

一是外感六淫。自然界因气候变化而产生的风、寒、暑、湿、燥、火六气是万物及人类生长的条件,但是六气变化过于急骤突然,在人体正气不足、抵抗力下降时就会成为致病因素,侵入人体使之得病,这时的六气称为"六淫"。

二是内伤七情。通常人体的精神情绪可概括为七种状态,即喜、怒、忧、思、悲、恐、惊。一般的精神状态变化不会使人得病,而突然的、强烈的、长期的情绪刺激,超过了人体正常的承受能力,会使人体气机紊乱,脏腑的阴阳气血失调,从而导致疾病发生。

三是内外杂合。诸如饮食失调、房事失节、劳逸失度、外伤失防等因素都能够影响人体的生理功能,使正气受伤,脏腑失调,以致引发疾病。

虽然致病因素有三大类,但从总体上说,疾病的预防不外乎两个方面。这就是体内与体外的预防。《素问·上古天真论》说得好:

"虚邪贼风,避之有时,……精神内守,病安从来。"

所谓"虚邪贼风,避之有时",是强调对外界一切容易致病的邪气,要注意适时回避,以防体外致病因素的侵袭。基本途径是,效法天地变化的规律,注意根据四季气候变化的特点,适时调养精气,防止"六淫"侵袭。人们常说的"秋冻春捂"、"冬要温床暖室,夏要净几明窗"、"凡人卧,春夏向东,秋冬向西"等民谚,都是防止外邪侵入的有效方法。

所谓"精神内守,病安从来",是要求保持体内的真气和顺,精神内守而不耗散,增强人体内在的防病机制。基本的途径是,思想上保持清静,无欲无求,饮食有一定节制,既不妄事操劳,又避免过度房事,防止七情失调和外伤失防。古人实行的精神养生、房室养生、气功养生、运动养生都是切实可行的防病途径。

内防也好,外防也罢,都是就人们自身如何防病而言的。而对于医生来说,还应注意运用药物预防疾病的发生或漫延。我国古代药物防病的方法林林总总,不可胜数,其中预防天花病的方法效果颇为明显,堪称中国人对人类预防保健的一项卓越贡献。天花是一种由天花病毒所引起的烈性传染病。中国本来没有天花,大约是公元2世纪时由南方传入中国的,此后即由南向北蔓延,为患深远,以致满清入关称帝后,曾因天花流行而惊恐,甚至为其继承人的安危而担忧。这种状态促使古代的医学家们长期致力于研究预防天花病的方法。早在公元8世纪时,就有医家采用天花病人的痂皮,以"吹鼻"、或种"人痘"的方法,使人感染上轻度天花而获得免疫来预防天花,取得良好效果。自公元17世纪起,这种接种人痘的方法开始向国外传播。公元1796年英国一位名叫琴那的医学家受接种人痘的启发,创造了接种牛痘的方法,并迅速得到推广。

1979年10月26日世界卫生组织在肯尼亚内罗毕宣布全球消灭了天花病。这光辉的一页既应归功于英国的琴那,更应归功于中国人发明的人痘接种术。既然天花可以预防,其它传染性疾病也可以预防。近几十年来的事实证明,斑疹伤寒、回归热、黑热病、麻疹、百喉、脊髓灰质炎及血吸虫病等一度死亡率很高的传染病,通过药物预防也已得到不同程度的控制。

中医对危害人类生命和健康的癌症、心脑血管病、高血压病,也已研制出一系列中药和医疗器械,如新发明的磁疗中药鞋,对部分高血压患者,具有一定的降压和镇静的作用。在《黄帝内经》中曾提到'重履而步',唐代初年医学家杨上善便认为'重履'即是将磁石置于鞋中,具有防治疾病的作用。

中医药预防保健方法十分丰富,而且多有简便易行的特点。如用艾灸足三里穴(膝下外侧三寸处)可以改善肠胃功能,增强体质,所谓"要想身体安,三里永不干。"

野菊花沏水代茶,可以清头明目,春天最宜饮用。用滑石六份、生甘草一份(称"六一散"),煮水服用,可以清热而预防中暑。

生姜红糖水可以散寒气,淋雨受寒后应饮一大碗,能预防感冒。秋天气候干燥,容易引起咳嗽,秋梨膏和二冬膏,都有润肺止咳功效;莲藕合冰糖煮水饮用,也能润肺,对咳嗽痰中带血的病人有治疗作用。羊肉是热性食品,冬天吃涮羊肉可以御寒,而夏天吃就不太适宜了。

紫苏叶、生姜能解鱼蟹毒,吃水产品时最宜配合用之,鲜姜与鲜苏叶可以洗净生吃。炒枣仁同橘皮煮水喝,可以健胃安眠。蜂蜜煮山楂，既能润便，又有一定的柔化血管作用，老年便秘之人，食之最宜。

用生苡米煮粥，或将生苡米配合大米煮粥，对慢性肾炎和冠心病患者，具有保健和一定的治疗作用。当归、生姜、羊肉同煮熬汤饮用，是治疗妇女因虚寒而月经崩漏不止的经典药方。

枸杞子泡酒，久服有补肾功效，能治肾虚阳萎，不过年轻火旺之人不应饮用，所谓"离家千里，勿食枸杞。"

即使不离开家眷，年轻体壮人服此药酒，也难免出现头晕、目赤、耳鸣等症状。"药食同源"，调配得当都可达到防病强身的目的。

显而易见，防与不防是大不一样的。未病先防，对于个人来说，可以抵御疾病的侵袭，增进健康，延年益寿；对于国家来说，可以增强国民的体魄，减少疾病的流传，乃至于提高生产力。而且，与"渴而穿井，斗而铸兵"相比，未病先防显然容易得多，有效得多，可谓事半而功倍。在某种程度上来看，防病比防敌更加困难。敌人往往是明确的，疾病却往往看不见摸不着，需要更加耐心、细致地防备，切忌事不大而不为，害不大而不防。

## 4、兵家重防微,医家重杜渐

"微"就是微小。然而对待微小的不同态度常常成为智与愚之间的分水岭。早在《孙子兵法》产生之前,被誉为兵家鼻祖的姜尚在其《太公兵法·金匮》中就指出:"道自微而生,祸自微而成。"看到了"微"是祸端,不可低估。历代智谋之士无不重视防微杜渐。因为"禁微则易,救末者难"(《后汉书·丁鸿传》)。这是无数历史教训的总结。唐玄宗忽略了这一点,结果导致了"安史之乱"。

"安史之乱",是唐朝中期安禄山、史思明发动的叛乱。其实,早在安禄山任平卢兵马使时就有人指出他有谋反之心,但是唐玄宗不以为意,反而被安禄山的甜言蜜语所迷惑,竟然使他同时兼任三镇节度使,掌握十五万大军。这真是养虎遗患。所以,公元755年安禄山伙同史思明在范阳起兵谋反时,唐玄宗仍然沉漫在轻歌曼舞之中,完全没有戒备,以致叛军势如破竹,一举攻破洛阳。虽然经过郭子仪等将领长达八年的奋力反击,终于平定了叛乱,但是大唐江山也因此而大伤元气,一蹶不振。这其中的教训是非常深刻的,而且是有广泛借鉴意义的。

在医学上,中国传统的医家们很早就认识到,病邪通常是由表入里的,如不及时治疗,就会坐失良机,疾病日趋深重,恢复也就不易。例如,感冒表面上看起来是小毛病,不少人不予重视,不加任何调治,以为抗几天就会好起来。殊不知,肺炎、气管炎、肾炎、心肌炎、风湿病等等,往往就是由这小小的感冒进一步发展而成的。不但外感病是这样,脏腑有病也是这样。脏腑有病也会相互传变。肝病会影响到脾,脾病又会影响其它脏器。所以,《黄帝内经》指出:"善治者治皮毛,其次治肌肤,其次治筋脉,其次治六腑,其次治五脏。治五脏者,半死半生也。"治五脏者犹攻城,系不得已而为之。这一主张的核心是要求对待疾病除了平时注意预防外,还要进行早期治疗,这样才能在病情容易治疗的时候一劳永逸,或者防止病情恶化。反之,则危害至深,难以治愈。

中国古代有一个"讳疾忌医,病入膏肓"的故事,形象地说明了防微杜渐,及时治疗的重要性。

《史记·扁鹊传》记载说,战国时名医扁鹊,有一天去见齐桓侯。他发现齐桓侯气色不太好,直言说:"您有病!这病在皮肤里,现在还不严重,如果不及时治疗,恐怕就要恶化。"齐桓侯冷冷地说:"我没有病。"扁鹊走后,齐桓侯很不高兴地说:"做医生的总喜欢把没病的人平白地说成有病,以显示他的医术高明。"

过了五天,扁鹊又去见齐桓侯,严肃地说:"您的病进入血脉了,若不赶快治疗,就会更严重。"齐桓侯还是不信。

又过了五天,扁鹊再去见齐桓侯,惊叫道:"您的病已经深入到肠胃了,再不治,就有危险了!"齐桓侯完全置之不理。

又过了五天,扁鹊一见齐桓侯,话也不说,转身就走。齐桓侯反而觉得奇怪,便派人向扁鹊问个究竟。扁鹊说:"病在皮肤,是容易治的,用热毛巾敷一敷就行了;病在血脉,也不难治,可以用针灸的办法;病到了肠胃的时候,也还有办法,吃几服汤药,仍有治好的希望;然而病入骨髓以后,就什么办法也没有了。现在桓侯的病就是已经到了骨髓里了。"

五天以后,齐桓侯果然起不了床,病情十分严重。急忙派人请扁鹊,扁鹊早已不知去向。不久,齐桓侯就病死了。

这个故事是有典型意义的,发人深省。汉代医学家张仲景就十分推崇这个故事,并在《伤寒论》中作了深入的分析。他说,"通常疾病刚刚侵入人体时,风寒既浅,气血脏腑尚未受到伤害,及时治疗自然比较容易。如果任凭邪气深入,则邪气影响体内的正气,并与之混合在一起。这时,欲医治邪气则妨碍正气,欲扶持正气则帮助邪气,即使邪气逐渐消失但正气也已经受到伤害。如果得病之后,不加注意,继续操劳,使病情加重,以致病上加病,那就尤其危险。"他告诫人们,如果稍有不适,必须及时调治,切不可象齐桓公那样讳疾忌医,以为病小而不加重视,以致逐渐深入;更不可勉强支持,使病加重,造成无法医治的后果。这是人们必须高度重视的问题,而医生则更应当注意对病人进行早期治疗,将病情遏制在萌芽状态。

在医疗条件有了很大发展的今天,防微杜渐的思想进一步丰富。人们不仅强调早期治疗,而且进一步强调早期发现,早期诊断,简称为"三早"。许多事实证明,只要做到"三早",即使对令人谈之色变的"不治之症"——癌症,也可以取得良好的治疗效果。而定期进行体格普查及疾病筛检,则既可使一些肺结核、肿瘤、肝炎等疾病的患者得到及时治疗,缩短疾病过程,提高疗效和减少费用,同时还有利于防止疾病蔓延。

## 5、兵家尚武德,修身重养性

中华民族从来就重视道德教化,积极的入世精神,优良的道德品格,几千年来如薪传火,连绵不绝,渗透到社会生活的各个方面。

毫无例外,道德素养在军事领域中,长期以来同样被置于首要地位,成为兵家思想的一种精神力量。《孙子兵法》关于将帅修养提出了"智信仁勇严"五条标准,又从反面提出了"将有五危",即"必死,可杀也;必生,可虏也;忿速,可侮也;廉洁,可辱也;爱民,可烦也。"意思是说,将帅有五种致命的弱点:只知死拚可能被诱杀,贪生怕死可能被俘虏,急躁易怒可能中敌人轻侮的奸计,廉洁好名可能入敌人污辱的圈套,一味"爱民"可能导致烦扰而不得安宁。《孙膑兵法·将义》又进一步发挥说:"将者不可以无德,无德则无力,无力则三军之利不得。故德者,兵之手也。"《吴子》则把将帅的"五德"概括为"理、备、果、戒、约。"总之,优良的品德是为将的首要标准。

养生必须修德养性,医学家们从对社会生活的考察中发现了修德养性对增进健康的重要作用。早在《黄帝内经》的首篇《上古天真论》中就指出了养性乃是养生长寿之道。

在中医看来,"性"是一个人品德的深层表现。只有修养心性才能驾驭情感,培养出一种心平气和、谦虚谨慎、随份从时、清静寡欲的道德情操,这样就能增强人体对外界刺激的适应范围和强度,具有抵抗疾病的能力。相反,必然使人的精神气血衰败,容易产生各种严重的疾病。

正如《黄帝内经·素问·汤液醪醴论》所说:"嗜欲无穷,而忧患不止,精气弛坏,荣涩卫除,故神去之而病不愈也。"若酒色财气追求无度,患得患失终日不宁,必将损害健康,既病之后若仍不知节制,其病将难痊愈。

唐朝医学家孙思邈《千金要方·养性论》指出:"夫养性者,所以习以成性,性自为善,""性既自善,内外百病皆不恶生,祸乱灾害亦无由作,此养生之大经也。"

要做到既不招灾又不惹祸,关键是要做到与世无争,豁达大度。反对争名夺利,蝇营狗盗,所谓"名利杀人,甚于戈矛"(《无字真经养真篇》)。

对于世事,明朝王象晋说得好:"大事难事看担当,逆境顺境看襟度,临喜临怒看涵养,群行群止看见识"(《清寤斋心编》)。王象晋这四"看",核心是只有看破名利。同为明朝人的胡文焕,他有几句诗真可谓金石之言:"何必炼丹学驻颜,闹非朝市静非山。时人欲觅长生药,对境无心是大还"(《类修要诀》)。所谓"大还",乃是气功修炼术语,指达到一种恬淡虚无,返璞归真的境界。

吕岩《吕帝文集》记述了一个神仙吕洞宾的"三剑"故事,可以作为此种修养心性的注脚。这个故事说,吕洞宾听了人们误传他用飞剑杀人的传闻后,笑道:"慈悲者,佛也。仙犹佛耳,安有取人命乎?吾固有剑,盖异于彼:一断贪嗔,二断爱欲,三断烦恼。此其三剑耳。"

在这里,吕洞宾告诉人们,佛也好,仙也好,都是慈悲为怀,哪会去杀人呢?他说他确实有三把剑,但同人们传说的不同,一把用来斩断贪婪之心,一把用来斩断爱欲之心,一把用来斩断烦恼之心。当然,无论军事学还是中医学,其所强调的养性都不是单纯的"恬淡虚无"而心无所想和无所适事,而是要求首先做到"形劳而不倦"(《黄帝内经·素问·上古天真论》),即劳逸适度,亦弛亦张;其次是要求精神专一、正直,排除各种杂念的干扰。这样,即使工作、学习比较紧张,仍然可以使身体的各种生理功能保持正常,而达到健康长寿的目的。正如《黄帝内经·灵枢·本藏》所说:"志意和则精神专直,魂魄不散,悔怒不起,五脏不受邪矣。"孙思邈之所以能写出宏篇巨著《备急千金要方》、《千金翼方》,对中国医药学做出重大贡献,被后世尊称为"药王",正是与他善于修身养性,从而保持体魄健全、思维敏达分不开的。而他寿逾百岁,也为其倡导的养生之道增强了可信度。显然,中医的"养性"理论,实际是一个人生观的问题。一个人的精神寄托只要做到境界高远而不患得患失,大事清楚,小事"糊涂",与人为善,宽宏大量,必然能长命百岁。

# 6、地势兵之助,地利人增寿

人们生活在天地之间,无时无刻不与自然界发生着密切的关联。饶有趣味的是,人们在适应和选择生存环境时,竟与兵法中的某些原则暗自相合。

《孙子兵法》在论述安营扎寨应当利用"地利"的原则时指出:"凡军好高而恶下,贵阳而贱阴,养生而处实,军无百疾,是谓必胜。"对照中国历代环境养生理论,可以毫不夸张地说,这一主张不仅仅明确了驻军的基本要求,而且概括了环境养生的主要原则,即使在今天,现代养生学对于居处环境也不外乎这些基本要求。我们不妨逐条分析一下孙子提出的这25字原则:

"好高而恶下",意思是说,军队安营扎寨喜欢干燥的高地,避开潮湿的洼地。让我们看看养生学家对选择居室的标准是怎么说的。明代养生学家陈继儒在《养生肤语》中指出,人之居屋须"高朗干燥,斯无患矣。"道理很简单,干燥的高地往往空气新鲜,潮湿的洼地则难免腐秽恶臭。通常成年人每天必须吸入15立方米左右的空气。含氧、氮、阴离子丰富的新鲜空气是人体维持新陈代谢必不可少的重要因素,而混浊、污染的空气对人体健康的危害是十分严重的。许多呼吸系疾病,如支气管哮喘、慢性支气管炎,以及肺癌、胃癌、心肌梗塞等都与空气污染有关。所以,《老老恒言》中也指出,"卑湿之地不可居",认为居住楼房可以杜绝湿气的危害。

"贵阳而贱阴",意思是说,军队应驻扎在向阳之处,避开阴暗之地。其实,人们在工作和生活中,对客观环境的要求也希望居室"通彻阳光"(《延寿新法》)。因为,只有向阳之处才能阳光充沛,光线明亮;阴暗之地则阳光难至,视界狭窄。万物生长靠太阳,这是众所周知的道理。阳光不仅具有杀菌抗病、清洁空气、增高室温的作用,而且具有影响精神和心理的作用,明媚的阳光往往使人精神愉快,进而促进身体健康。所以,我国人民建造居室时历来注重采光,房前屋后虽植树木,却不使树荫遮住门窗。《老老恒言》认为,适于养生的居处应当"两旁空阔,则红日满窗。……室前庭院宽大,则举目开朗,怀抱亦畅。更须树荫疏布,明暗适宜,如太逼室,阳光少而阴气多。"

"养生而处实",意思是说,军队驻防的地方要靠近水草丰富,交通便利的地区。"背山临水,气候高爽,土地良沃,泉水清美"(《千金要方·择地》),也是杰出的医学家孙思邈选择居住的条件。这显然与孙子的主张有着异曲同工之妙。他自己正是在这样的环境中生活,一直活到102岁的高龄。

水,是人类赖以生存的必需物质,清洁和富含微量元素的水是人体健康和长寿不可缺少的。同时,水是万物之源。有水之处通常也是植物茂盛、风景秀丽之所,有益于人的身心健康。所以,人的居处既要接近于水源,又要交通便利。《起居安乐笺》中对居处的要求是,"使居有良田广宅,背山临流,沟池环市,竹木周布,场圃筑前,果园树后,舟车足以代步涉之难,"其中就兼顾了靠近水草和交通便利这两个因素。

如果能够同时满足这三方面的要求,无疑将有利于人体健康,延年益寿,古代善于养生的老寿星大多就居住在这种环境之中。由此,我们也不难理解,为什么那些企求长生不老的佛徒道士们要将寺庙道观建造在风景秀丽的名山大川,为什么那些希望万寿无疆的历代帝王要将行宫选定在依山傍水、四季如春的处所。

但是,人们生活的自然环境,有时往往难以随意选择,现代社会尤其如此。如在城镇工作的人,就难以选择山村作居处。因此,上述三条原则往往客观上难以同时实现。然而,人们可以主动改造环境,因地制宜,创造和美化居室周围的环境。在这方面,传说中的有巢氏早就为后人作出了榜样。《韩非子》中记载:"上古之世,人民少而禽兽众,人民不胜禽兽虫蛇,有圣人作,构木为巢以避群害,而民悦之,使王天下,号曰有巢氏。"尽管有巢氏构木为巢可能仅仅是出于防御野兽侵害及躲避风寒袭击的本能,客观上却开创了中华民族利用和改造居处环境以养生的先河。

既然上古之人都能够改造自然,营造居处环境,那么处于物质文明和精神文明高度发达的今天,人们无疑更有能力、更有条件改造和美化居处环境。一方面,可以在庭院之内,阳台上下,种树植花,使草木葱郁,鲜花常开。有条件的话,还可以建造一点假山假水,养鱼养鸟,营造一个山水兼备、百花争艳、枝叶青翠、错落有致的"袖珍花园"。这样既有利于净化空气,减缓噪音,又能够增加生活乐趣,锻炼身体。无庭院阳台的住户,可在室内养植几盆常绿花,如吊兰、仙人掌、君子兰等无香植物。它们不仅能净化室内空气,又能益于眼睛。在冰封大地的季节,常绿植物使室内充满着春天气息,有生发温馨之感。另一方面,设置南北门窗,经常开启,使室内光线充足,空气流通。夏季窗户虚敞,通风透凉;冬季定时开窗,保暖透气。并且经常打扫室内外的卫生,保持居处清洁,避免环境污染。如此则能防止疾病,有益于长寿。

## 7、兵非多益,食非过益

民以食为天。饮食是维持人体生命活动的基本条件,其重要性是自不待言的。但是,唐朝著名的医学家孙思邈却告诫人们:"万病横生,年命横夭,多由饮食之患"(《摄养枕中方》)。如果饮食不当就会招来各种疾病,甚至短命夭折。清代医学家徐大椿则以用兵为喻,形象地说明饮食不当的危害。他在《医学源流论》中说:"古人好服食者,必生奇疾,犹之好战胜者,必有奇殃。"

那么什么是饮食不当呢?简单地说,就是"过"。喝得过多,吃得过饱,粮食过细,肉食过腻,诸如此类都是"过"的表现。

现在,人们生活水平提高了,那种为了填饱肚子而过食、强食的现象已不多见,但是因为美味佳肴而大吃大喝的事却是相当普遍的。殊不知"五味之过,疾病蜂起"(元·朱震亨《格致余论》)。因为鸡、鸭、鱼、肉、精米细粮经煎炸、熏烤、油炒、火炖之后,如果过多摄取,就容易因内热过重而引发热毒、疮疡、痰热等病证和消渴、痈肿等疾患。这与现代医学提出多食这类食物,可以导致或容易引发高脂血症、肥胖症、糖尿病、高血压、动脉硬化、冠心病、痛风疖肿感染等疾病是一致的。

偏食是过食的"怪胎"。偏食往往造成体内营养素不适当或不平衡,引起机体代谢紊乱,细胞免疫功能下降,因而对某些感染抵抗力下降,而发生多种疾病。

古典医书中说:"多食咸,则脉凝泣而色变;多食苦,则皮槁而毛拔;多食辛,则筋急而爪枯;多食酸,则肉胝皱而唇揭;多食甘,则骨痛而发落。"这些都是告诉人们偏食对人体是有害的。

那么,怎样饮食才有益于养身防病呢?

早在二千多年前,孙子就提出"兵非多益",认为要取得战争胜利并不一定兵众愈多愈好。《黄帝内经》吸收了这一思想,看到饮食过度的危害,主张食非过益,贵在能节,因而在总结长寿秘诀时,将"饮食有节"作为重要的一条。

**所谓"饮食有节"主要包括五方面内容:**

一、饮食适量,要有节制,勿过饱、勿过味、勿偏食。合理安排膳食,配调适宜,主食要注意粗细粮混合,副食最好荤素搭配,做到"谨和五味"。尤其是老年人,阴虚火旺,消化功能衰弱,容易饥饿,但稍多食后又消化不了,反而引起食欲减退,所以必须按照规律,在饮食的数量上适当控制。常言说得好:"先饥而食,先渴而饮;食欲数而少,不欲顿而多,""晚餐少一口,能活九十九。"

二、饮食要有规律,宜早食,忌夜饮,不要饱一顿、饿一顿。我国人民一般一日三餐,因为一般性食物的消化吸收至少需要 4~5 小时,故早餐宜在早晨 7 点前后,午餐宜在中午 12 点前后,晚餐宜在下午 6 点前后。

三、适当忌口,对有些不需要的或对身体有不利影响的食物应当少吃,或者不吃。元代名医朱丹溪曾指出:"好酒腻肉,湿面油汁,烧炙煨炒,辛辣甜滑,皆在所忌。"有些医书还强调"生冷瓜果要少吃,免得秋来生疟痢。"

四、饮食要得法,注意暖、缓、软。所谓"暖",是指脾胃喜暖而恶寒,所以无论冬夏,饮食都应暖,少吃生冷之物,以免损伤脾胃。所谓"缓",是指饮食时应细嚼慢咽,切忌狼吞虎咽,尽量让其精华吸取,以便滋养五脏。所谓"软",是指坚硬之食最难消化,所以饮食宜软化,特别是"老人之食,大抵宜温热熟软",有利于脾胃运化以取得水谷精微而营养全身。宋代大诗人陆游晚年对此特别重视,专门作《食粥》诗一首。诗中写道:"世人个个学长年,不悟长年在目前。我得宛丘平易法,只将食粥致神仙。"

五、饮食要卫生,既要注意食物的新鲜、清洁,又要养成良好饮食习惯。对于食物要有所选择,注意"浆老而饭馊不可食"、"生料色臭不可用"、"猪、羊疫死者不可食"、"诸果落地者不可食",以免病菌由口而入。据说,唐代大诗人杜甫就是因此而猝死的。《明皇杂录》记载:"杜甫客耒阳,游岳祠,大水遽至,涉旬不得食。县令令具舟迎之。令尝馈牛炙白酒,……甫饮过多,一夕而卒。"郭沫若认为是嗜食县令所赐腐败牛肉,引起胰胆功能急性病变而造成死亡。

此外,对于饮食习惯也要有所节制。历代寿星的经验是"食勿大言"、"饱食不得急行"、"饱食不得便卧"、"食毕摩腹能除百病"、"食毕行步踟蹰则长生"等等,如此才能充分吸取和消化所食之物,避免因消化不良所引起的各种疾病。

清朝人赵翼在其所著《檐曝杂记》中记载了这样一个故事:有一位老寿星九十多岁了,有人问他长寿的秘诀是什么,他回答说:"好吃的不多吃。"他的回答对于那些为口伤身的人来说,无疑是一种警策之言。

由于我国人民几千年来十分重视"饮食有节",所以长久以来大多以素食为主,而西方人大多以动物性饮食为主,且甜食较多。正因为饮食结构的不同,高脂血症、肥胖症、糖尿病、高血压、动脉硬化、冠心病、痛风疖肿感染等疾病的发病率和死亡率,西方国家都明显高于中国。这个事实证明,我国传统的饮食养生理论具有无可置疑的科学性。

## 8、养兵莫贵于习练，养生莫善于习动

报载：94岁高龄的陈立夫先生耳聪目明，思维敏捷。有人问他养生之道，他以八字回答："养身在动，养心在静。"陈老先生可谓深得中医学的养生之道。养身与养心都是养生，但是在这一节里，我们单说运动健身问题。

古代医学家们认识到"药补不如食补，食补不如锻炼"，提出了"养生莫善于习动"的观点。战国时期，《吕氏春秋》一书中明确指出了运动养生的原理："流水不腐，户枢不蠹，动也。形体亦然。形不动则精不流，精不流则气郁。"

常言说，活动、活动。要想活就要动。动，才有生命力。其实，天下万事万物，莫不如此。俗语讲："拳不离手，曲不离口。"这是说拳师、演员必须经常练习自己专业的本领，才能具有旺盛的艺术生命。同此道理，军事上更加注重训练。《将苑》说："军无习练，百不当一；习而用之，一可当百。"诸葛亮的这段话，无非是强调养兵的核心在于练兵，养而不练，等于白养。从这个意义上讲，练兵与练身是可以相互为喻，说明其重要性的。

在日常生活中，反对运动健身的人几乎是不存在的。大量的人虽然懂得"生命在于运动"的道理，但由于种种原因不能坚持不懈，一以贯之。陈立夫先生则不然，他说："我每天清晨5时半一起身，就要淋浴，当水冲到哪里，就按摩到哪里，从头顶到脚心，每处用两手按摩一百下，一共需要40分钟。早饭后，再散步一千步。如此做法，已有28年，绝没有一天间断，此之谓'养身在动'。"健身的方式多种多样，难就难在像陈老那样"28年，绝没有一天间断。"

"夫神大用则竭,形大劳则敝"(《史记·太史公自序》)。在运动养生方面的另一种倾向就是运动量过大,那也会适得其反。《庄子·达生》讲了这么一个故事,哲理隽永,发人深思。

庄子说:有一个叫东野稷的人,驾驭马车的技术十分高明。他在卫庄公面前表演驾车的前进后退,左右回旋,令人叹为观止。庄公称赞说,即使是古代驾车能手造父再生也超不过他。并让他再去兜一个圈,表演给大家看。东野稷为了显示自己的本领,继续表演。结果马因精疲力尽而累垮了。

如此看来,选择适当的运动方式是养生的重要途径之一。特别对于老年人,更应"量体裁衣",劳逸适度。锻炼必须循序渐进,宜从小运动量开始,中运动量为止;动作以缓慢、协调、简便为好;运动后感觉全身发热、无疲劳感、食欲增加、睡眠良好为宜。

为着老年朋友考虑,在这里我们要着重谈谈舞蹈。舞蹈似乎是近年来人们文化生活水平提高后才兴起的健体养生方法。其实,早在上古时代人们就已经知道通过舞蹈来健体祛病。当然,那时的舞蹈不象现在温文尔雅的交际舞,而是模仿各种动物的特点,边嚎叫、边蹦跳。《吕氏春秋·古乐》中记载,"昔陶唐氏之始,……民气郁阏而滞着,筋骨瑟缩而不达,故作为舞以宣导之"。人们在患病之后,常常以舞蹈娱悦神灵,乞求保护。由于歌舞本身那种和谐的形体运动调理了体内的机能,使精气畅通,从而驱除了病魔,自己医治了自己。1973年考古学家在湖南长沙马王堆发掘一座汉墓,出土了大量医书,其中一幅宽50厘米,长100厘米细笔彩绘的《导引图》颇引人注意。画面4层,每层绘11人形,共有44个演练各种舞蹈动作的人物图像。每一图旁都注有文字,说明可以治某病症等,不少动作是模仿禽兽的飞翔、寻食、奔走的形态特点。这说明,至少到汉代,舞蹈已是一种相当普遍的健身方法。三国时,华佗使舞蹈健身的方法进一步规范化。他继承前人的经验,根据虎、鹿、熊、猿、鸟的特点,编制了一套"五禽之戏"。如果"体有不快,起作一禽一戏,怡而汗出,因以著粉,身体轻便而欲食"。文学家傅毅在《舞赋》中盛赞舞蹈可以"娱神遗老",是"永年之术"。现今伴随优美的音乐跳交谊舞,可以使人感到心旷神怡,精神愉快,增强食欲,有益睡眠和身心健康。

时代不同了,如今的生活方式比之于古人已经发生了巨大的变化,出现了各种各样的健身方法,创造了形形色色的健身器材。无论什么方法,何种器材,只要持之以恒,动而适度,就能达到娱悦身心,健体养生的目的。

## 9、一张一弛是文武之道,劳逸适度为养生之宝

治身如治国,晋人葛洪在《抱朴子内篇·地真》中曾经作了形象的说明,他说,"神"好象国君,"气"好象人民,"血"好象群臣。因此他主张"夫爱其民所以安其国,养其气所以全其身。"抱朴子这一深富宏观整体思维的养生理论,得到了后世养生家的好评。其实,抱朴子这一"治国如治身"的养生理论的本源却是出自诸葛亮。诸葛亮,人们都很熟悉。他是我国历史上一位著名的军事家,通晓兵法,足智多谋。但是他对养生学也有研究一事就鲜为人知了。他曾说:"夫治国犹如治身,治身之道,务在养神;治国之道,务在举贤,是以养神求生,举贤求安。"在诸葛亮看来,智士仁人是安邦治国的根本,养神蓄气是延年益寿的关键。

可是诸葛亮自己却没有长寿,54岁就病死了。关于诸葛亮之死因,他的对手司马懿可谓一语中的,那就是"食少事繁,焉能久乎!"抛开诸葛亮鞠躬尽瘁的精神不谈,纯从养生学的角度看,他积劳成疾,气血亏耗,必然加剧衰老和死亡。

常言说,不会休息的人就不会工作。工作和休息,劳和逸,既相互对立,又相互统一。

《黄帝内经》很早就认识到了人体生命活动的这种基本形式,提出"五伤"之说,即"久视伤血,久卧伤气,久坐伤肉,久立伤骨,久行伤筋"。

"久"就是"过",超过限度。无论是脑力劳动还是体力劳动都必须劳逸适度。明白这些道理是容易的,实行这些道理就不那么容易了。薛福辰是一位精通医术的名医,曾因入宫给慈禧太后治

病而名噪一时,但是,他只活了58岁就病死了。什么原因呢?"用心过度,内受损而不自知。"原来薛福辰在事业上"用心本专",已够伤神耗精了,可是他在医务之余,又嗜棋如命,常常"秉烛达旦,或演棋谱,或与客对奕"(《庸庵文别集》卷六)。他的前妻王夫人因为他下棋无度又不听劝阻,气冲冲将围棋扔到井里。可是,王夫人死得早,而薛福辰又旧习不改,过分贪棋,以致耗尽精力而早逝。

在从事脑力劳动的知识分子当中,不少人却是因为在事业上用心过度而损寿伤身。李白怜悯杜甫写道:"借问如何太瘦生,总为从前作诗苦。"杜甫未能长寿,当然还有生计艰难的原因,但他过分执着于诗歌的创作不能不说是原因之一。杜甫等文人是因为没有条件去追求养怡之福,那么对于有条件的人来说是不应当忽视的。宋人倪思对此算是看得真切了,他说:"造物劳我以生,逸我以老。少年不勤,是不知劳也;老年奔竞,是不知逸也。天命我逸,而我自劳,以取困辱,岂非逆天乎"(《金钼堂杂志》)?

其次,知逸切勿过逸,"过逸成恙"。一个人如果过于安逸、闲散,长期不从事劳动,会影响体内气血的正常活动,产生一系列病理变化而导致"逸伤"。"常欲小劳,但莫大疲及强所不能耳。"这是唐代医学家孙思邈在《千金要方》中提出的主张。

"小劳"既指体力劳动,也指脑力劳动。无论是否知识分子,经常用脑是有益于健康的。俗话说:"常用脑,可防老。"古今中外,大凡长期从事脑力劳动的人一般都长寿。我国科学工作者曾对秦汉以来至本世纪四十年代末的三千多位科学家、艺术家、文学家和思想家的寿命进行了统计,算出其平均寿命是65.18岁,比一般人的平均寿命高出30岁左右。国外也有学者对16世纪后欧美的人口作过分析,那些平时用脑最勤的发明家、科学家的平均年龄为79岁,如爱迪生84岁、伽利略78岁、牛顿85岁、达尔文73岁、爱因斯坦76岁、福克雷114岁、罗素98岁等等。现代科学研究证明,一个人经常用脑,不但不会加速衰老,反而能够延缓衰老,防止老年性痴呆。

脑力与体力要结合,要坚持体育运动或体力劳动,并且从年青时就要打好基础。南宋爱国主义诗人陆游一生坎坷,却能高寿享年85岁,是与他平时注重养生健体分不开的。他曾自称"六十年间万首诗",平均两天要作一首,如此频繁而紧张的脑力劳动,如果没有强壮的体魄是难以持久的。所以,陆游十分注意养生,除注意起居养生、精神养生和饮食养生外,在运动养生上也颇下功夫。

他从小就热爱体育活动,喜欢打球,青年时期是一员球场健将。他诗中记述道,"军中罢战壮士闲,细草坪郊恣驰逐。洮州骏马金络头,梁州球场日打球",生动地描写了郊外打球的场景。晚年回乡之后,他勤务农事,以此调理身体,并把参加田间劳动和读诗、写诗联系起来,创作了大量具有田园气息的诗歌。如"小园烟草接邻家,桑柘阴阴一径斜,卧读陶诗未终卷,又乘微雨去锄瓜"。他还注意坚持每日打扫环境卫生,把一些小劳小动也当作健身之法,"一帚常在旁,有暇即扫地。既省课童奴,亦以平血气。按摩与导引,虽善亦多事,不如扫地法,延年直差易。"

现代社会中,人们的生活节奏空前加快。要想适应社会的要求,我们不能不发扬诸葛亮鞠躬尽瘁的精神,努力学习新知识、掌握新技术,同时也应不妨取法于陆放翁,注重多途径养生,尤其是运动养生,以保证劳逸结合,身心健康。

## 10、怒而兴师必败,愠而处世戕身

喜怒哀乐,七情六欲,是人与生俱来的本能。人是情感动物,不可能没有情感。正常情感的表露,为人的一生编织了一个又一个色彩斑烂的花环;异常情感的表露,则为人的一生套上了一个又一个阴冷沉重的锁链。

"百病起于情,情轻病亦轻"(《击埌集》卷17《百病吟》)。这是宋人邵雍的至理名言。情有多种多样,中国人一般概括为七种,即喜、怒、忧、思、悲、惊、恐。在这七情之中,"怒"这种情感危害最大。明朝来知德说得好:"人之七情,唯怒难制"(《来瞿唐集》)。

怒气难制,不只是平常百姓制不了,大人物中制不了的也多的是。再说了,小人物之怒同大人物之怒后果不同。小人物怒火中烧,失去理智,充其量影响他自身的健康和家庭的幸福。大人物则不然,不仅伤身,更为重要的是祸及国家,殃及人民。孙子早就看到了这一严重性,他在《孙子兵法·火攻篇》中向国君和将帅大声疾呼,告诫他们要"制怒",特别是战争决策上要"制怒",在兴师动众上要制怒,所谓"主不可以怒而兴师,将不可以愠而致战。"怒就是愠,都是指人的激愤情绪。无论国君还是将帅,一旦处于这种心理状态,很难理智地分析问题和正确地处理问题。如果凭一时怒气而兴兵作战,则难以清楚地了解敌情我情,合理地排兵布阵,灵活地变化战略战术,往往很容易导致失败。

《三国演义》中火烧连营七百里的故事是家喻户晓的真实史例。公元219年末,孙权派兵偷袭荆州,擒杀了蜀将关羽。刘备万分悲痛,发誓要为关羽报仇。赵云劝刘备说:"国贼乃曹操非孙权也。今曹丕篡权,神人共怒,陛下可早图关中,屯兵渭河上流,以讨凶逆。则关东义士必裹粮策马以迎王师。若舍魏以伐吴,兵势一交,岂能骤解?愿陛下察之。"向刘备劝谏的大臣很多,但刘备激于私愤,丧失了理智,一心只想为关羽报仇,夺回荆州。他既未看到蜀吴力量对比达不到灭吴的目的,又忽略了魏蜀吴三角鼎立的战略格局中有一个螳螂捕蝉黄雀在后的利害关系问题。公元221年,刘备亲自率领八万大军进攻吴国。盛怒之下的刘备,求胜心切,没能冷静地分析敌情地形,结果被东吴将领陆逊火烧连营,损失了全部舟船器械和大部分兵力,从此蜀国一蹶不振。

刘备此举,不仅使蜀国丧失了元气,他自己也丧失了元气。据张从正在《九气感疾更相为治术》中指出:"怒气所至,为呕血,为飧泄,为煎厥,为阳厥,为胸满胁痛;食则气逆而不下,为喘渴,烦心,为消瘅,为肥气,为目暴盲、耳暴闭,筋解;发于外为疽痛。"以此看来,刘备在彝陵之战后很快就病死,当与张从正所说的种种症状相合。试想,刘备先是因东吴占荆州、杀关羽而怒火中烧,后又在彝陵惨败于一个无名小将陆逊的手中,更是丢尽了面子,威风扫地,由怒而哀,由哀而恐,整个身心都彻底崩溃了。年63岁一命呜呼,实在是咎由怒起。

"怒"的危害是严重的,"怒"对人生是可怕的,古人对此不胜感慨。大诗人陆游在《杂兴》一诗中写道:

> 灵府宁容一物侵,
> 此身只合老山林。
> 何由挽得银河水,
> 净洗群生忿欲心。

陆游想用与世无争的态度以去欲平怒。

明朝胡文焕在《类修要诀》中写了一首著名的《戒怒歌》,歌词云:

君不见：
大怒冲天贯牛斗，
擎拳嚼齿怒双眸。
兵戈水火亦不畏，
暗伤性命君知否？

又不见：
楚霸王周公瑾，匹马乌江空自刎。
只因一气殒天年，空使英雄千载忿。
劝时人，须戒性，纵使闹中还取静。
假若一怒不忘躯，亦至血衰生百病。
耳欲聋又伤眼，谁知怒气伤肝胆。
血气方刚宜慎之，莫使临危悔时晚。

胡文焕的《戒怒歌》显然比陆游的《杂兴》诗进了一大步。他不仅指出怒气对事业的影响，更从传统医学的角度说明了怒的危害和制怒的方法。

传统医学认为，怒伤肝。因为当人发怒时，肝气上逆，肝举叶张，迫使气血上冲，而致头胀眩晕，面赤，头痛，甚至昏仆。

现代医学的一些研究资料也证明，在平时比较容易激动、喜怒无常的人群中，高血压病、心脏病、溃疡病、甲状腺机能亢进等疾病的发病率为77·3%，比性格乐观的人的发病率高出好几倍。

现代医学进一步深入揭示了其中的缘故。一个人如果长期处于过分紧张、焦虑的情绪中，会引起心律失常、心动过速、心前区疼痛等症状。而盛怒之中的人则常常血压升高，如果经常发怒，使血压反复升高，就会导致高血压病。长期过度的思虑和抑郁容易引发溃疡病；不良的精神刺激会引起糖尿病、甲状腺机能亢进或减退；平常过于抑郁寡欢的妇女会出现月经不调、乳汁减少等症状。

既然以愠处世必将导致严重疾病，那么要想健康长寿就不能不注意研究控制自我情绪的艺术。清代梁章钜在《退庵随笔》中推荐了一个"独守方寸之地"的方法，也就是稳定心绪，安定精神的方法。他主张固守心境犹如禁闭心中的城门，把东南西北四门关闭得严严实实，一丝一毫也不容荣辱、进退、升沉、劳苦、生死、得失等一切可能引起情绪波动的意念进入其中。如果偶有不慎，使得某种感情杂念打进城来，便要立即加以驱逐。其核心是效仿道家"无为"、"无欲"的主张。

曾国藩说得更透彻，认为恼怒如同蝮蛇，戒怒制怒去怒要有非凡的勇气。他说"此病非药饵所能为力，必须将万事看空，毋恼毋怒，乃可渐渐减轻。蝮蛇螫手，则壮士断其手，所以全生也"（《曾文正公家书·致沅弟书》）。

《黄帝内经》指出，"怒则气上"。常言所谓怒发冲冠，头发都竖起来了。产生恼怒的根源既然在气，那就要养气养心，保持心平气和的心境，"大怒不怒，大喜不喜。"

这些方法是值得借鉴的，它对于控制情绪无疑有一定的效果。当然，今天的人们不可能像古人那样完全处于"万事不关心"的静守状态，但每天利用一定的时间练练气功、培养元气，却是十分有益的。最有效的办法恐怕还在于学会主动地掌握自己的情绪，适时调和七情，使自己的情绪无论在什么情况下都能够处于怡情悦性的状态。"生气催人老，笑笑变人少"，"笑一笑，十年少；愁一愁，白了头"，这些流传久远的民谚仍可作为我们调养情志，益寿延年的座右铭。

## 11、心静利于制胜,情怡益于永年

在中国成语中,与"心"字有关的实在不少,诸如"心平气和"、"心甜意洽"、"心旷神怡"、"心明眼亮"、"心劳日拙"、"心惊肉跳"等等,不下数十条。中国人重视"心"已有两千多年的历史。那时,人们误以为"心"是管思想意识的,是"五脏六腑之主",相当于现代观念的大脑。后来,人们虽然知道了人的意识思想与大脑的关系,但"心"这个词一直相沿成习,变为思维的同义语。

养生家把养心放在极其重要的地位,而求取心静是养生的基本方法,是获得生命功能的源泉。正如《老子》所说:"夫物芸芸,各复归其根,归根曰静,是谓复命。""静"是生命的根,生命的常,只有"守静笃",才能生命长在而无危险。南北朝时的医学家陶弘景进一步指出,"静者寿,躁者夭。"《老老恒言·燕居》中也有"养静为摄生首务"之语。

对于虚静的追求不仅是为了养生,也是为了养性。不只是养生家重视,诸子百家也无不重视。《孙子兵法》有一句名言:"将军之事,静以幽,正以治。"这一名言,古今中外的军事家无不奉为圭臬。

静、幽、正、治是孙子对统兵将帅进行性命双修提出的四项要求。在孙子看来,作为一个优秀的将帅,要想获得镇定、沉着、临危不惧的品质,必须在心理上做到虚静、深沉,物我两忘。《三国演义》中,诸葛亮敢于采用空城计,就表现出非凡的镇静。小说中写道:蜀将马谡因刚愎自用而失守街亭,致使魏将司马懿的十五万大军直逼西城。当时驻军西城的诸葛亮只有一班文官和二千五百名士卒,来不及调兵遣将。在众官尽皆失色之际,诸葛亮沉着冷静,命令所有军民都隐蔽起来,大开城门,仅派几十名士兵扮作百姓在城门附近洒扫街道,自己则端坐于城头悠闲自得地弹奏瑶琴。司马懿来到城下时,以为城中埋伏有重兵,吓得马上转身退走。这个故事虽是小说家言,但可称得上是对孙子上述名言语的最好注脚。

无论是将帅还是文臣,是官吏还是平民,要真正达到清静无为的境界是必须悉心修养、刻苦修炼的。略而言之,有以下几点:

**首先,要注意培养品德**。传统医学主张性命双修,从来是把养德与养生相提并论的。他们认为,讲道德,重仁义,不谋私利,不患得患失,既有利于思想境界的升华,又有益于心志安定,气血和调,从而使人体的生理活动能够按正常规律进行,精神饱满,形体健壮。先秦时期的大思想家孔子在《中庸》中提出,"大德必得其寿",即有高尚道德修养的人,可获得高寿。西汉前期的董仲舒,将养心与中庸思想结合起来,认为"能以中和养其身者,其寿极命"。唐代孙思邈在《千金要方》中则从医学角度指出:"夫养性者,所以习以成性,性自为善。……性既自善,内外百病皆不悉生,祸乱灾害亦无由作,此养生之大径也。"总之,只有在养生中注重养德的人,才有可能"跻仁寿之域。"

**其次,要能够"恬淡虚无"**。人们生活在大千世界之中难免受到这种或那种矛盾的困扰,诸如失恋、丧亲、事业不顺等等,这些矛盾都可能引起情志上喜、怒、哀、乐的变化,这是一种必然的心理反应。一个人如果对于这些事情看得过重,则会长期沉浸在不正常的情绪之中,或终日愁眉苦脸,或时常唉声叹气,或突发莫名怒火,久而久之势必成为导致心理疾病的因素,乃致损寿折年。对此,《黄帝内经》提出了一剂良方:"恬淡虚无,真气从之,精神内守,病安从来。"意思是说,在思想上要淡薄名利,保持清静安宁,不贪欲妄想，就可以保持精神健旺，抵御疾病的侵袭。传说中的彭祖，是我国历史上最著名的寿星，相传活到了800岁，这虽然过于夸张，但长寿是无疑的。《彭祖摄生养性录》一书中记载了关于调养精神的秘诀，即"不思声色，不思胜负，不思曲折，不思得失，不思荣誉，心无烦。"古代流行的《孙真人卫生歌》唱得好："世人欲识卫生道，喜乐有常嗔怒少。心诚意正思虑除，顺理修身去烦恼。"总之，只有淡薄名利，泰然处世，知足常乐，才能调和七情，使之怡然融合，从而情绪安定，精神愉快，进而保证脏腑协调，气机顺畅，身体健康。

**再次，要善于自我调节**。为了实现清静，养生学有两条原则是值得重视的。一条是"我命在我"，自己掌握自己的命运，妥善处理烦乱的人事。对于大多数人来说，人生的旅途是坷坎不平的，事情十之八九不能尽如人意，人们一旦遇上这一类事，就应当善于控制自己的情绪，冷静处置，切忌惊慌失措，急躁从事。中医古书《友渔斋医话》中清楚地说明了处难不惊的心理效应，即"遇逆境，善自排解，因肾水上交于心，此者，病从何来？"心属火，肾属水，心肾相交，水火既济，自然健康无病。然而，自我调节往往不是心想事成的，必须在平时养成制怒、节哀、广思、怡乐的习惯，才能遇事平静，避免不良的精神刺激。另一条原则是"静在心，不在境"。只要人心能静，无论客观环境如何嘈杂纷纭也能做到澄然无事。反之，如果自己心绪不静，即使环境优雅，也不会求得安宁的心态。为了心静，为了境静，养生家不断摸索各种方法。

例如宋代养生家陈直在《养老奉亲书》中所提出的"五事"和"十乐"可谓概括了古人调和心情的基本方法。所谓"五事"，是指静坐第一，观书第二，看山水花木第三，与良朋讲论第四，教子弟第五。所谓"十乐"，是指读义理学，学法帖字，澄心静坐，益友清谈，小酌半醺，浇花种竹，听琴玩鹤,焚香煎茶,登城观山,寓意奕棋。

这其中包括了传统文化的大部分内容,从琴棋书画,到诗赋文章,从名山大川,到庭院梅竹,无不具有陶冶情趣,修身养性的作用,常年陶醉其中,自然可以使人进入恬淡虚无的境地,遇有不尽人意之事也就能泰然处之。除陶冶情趣之外,更重要的是必须培养出"泰山倾而不惧,黄河决而不乱"的心理素质。这就需要勇于面对现实,热爱生活,执着追求,无论是遇到大喜还是大悲,都能从容不迫。孙子要求将帅要能够"杂于利害",做到"杂于利,而务可信也;杂于害,而患可解也。"也就是说,聪明睿智的将帅必须善于兼顾利与害两个方面,在不利的情况下要看到有利的因素,则大事可成;在有利的情况下要看到不利的因素,祸患就有可能预先排除。平时养生若能如此,无疑有益于自我调节,稳定情绪,以致七情和悦,怡然处世,真正做到心如明镜,情如止水,延年益寿。

# 12、养兵须作息有序,养生当起居有常

俗话说:"养兵千日,用兵一时",它揭示出"养兵"与"用兵"的密切关系。"用兵一时"能否奏效,关键在于"养兵千日"是否得法。《孙子兵法》尽管是一部专讲"用兵之法"的兵书,但对于"养兵之法"也不乏高屋建瓴之见。其中一些观点不仅适于"养兵",也适于"养生"。例如,他在《九地》篇中几乎是以养生学家的口吻提出军队要"谨养而勿劳"的主张,要求军队平时应当严格注意休整,避免使将士过度疲劳。谨养勿劳,就是要保证军队有规律的生活和训练,即作息有序。因为,只有作息有序才可能使部队避免过于疲劳,或过于闲逸,保证将士得到充分的休养,从而达到提高士气、积蓄力量的目的。

练兵最难的在于养成教育。无论是平时还是战时,站有站像,坐有坐像,始终保持一种良好的军人姿态。这样的军队,就是训练有素的军队,就是能打胜仗的军队。既然孙子所说的"谨养"能使百万之众的军队获得无形的力量,那么有规律的日常生活对于一个人的健康无疑更为有益。《黄帝内经》开篇就对"度百岁乃去"和"半百而衰"的两类人作了比较,认为起居是否有规律是造成这种差别的重要原因之一,前者"起居有常",后者"起居无节"。

"起居有常",就是人们必须有规律地安排生活起居,养成良好的作息习惯。在中医学上,起居泛指生活作息以及平素的言谈举止、衣着打扮,时时处处从琐细的起居卫生入手,才能真正做到颐养天年。事实上,从古至今的老寿星大多都是生活最有规律的人。中国当代著名的文学家、教育家叶圣陶先生在人生旅途上度过了

94个春秋，堪称名副其实的老寿星。他的长寿秘诀之一就是生活有规律。他曾自述道："每天7点半早餐，12点午餐，6点过后晚餐，习以为常。""一般是睡眠时期极长，夜8点半睡，下一天早上5点半光景醒。"即使在星期天，他也总是坚持"上午看书，下午沐浴"，很少变化。正是因为生活规律，所以叶先生的身体几十年间一直保持良好的状态，很少生病。

那么，人体的规律是什么呢？以天人感应学说为理论基础的中医学认为，人是自然之子："人生于地，悬命于天，天地合气，命之曰人"，"人以天地之气生，四时之法成"（《黄帝内经》）。天地是人类的母体，生命的源泉，所以人的一切都必须"与天地相参，与日月相应。"也就是说，人类的一切生理现象及活动规律都与自然界的运动变化息息相关。例如，一年四季之中，气候有春温、夏热、秋凉、冬寒的变化，人体则表现为春夏阳气渐长，秋冬阴气渐旺，即阳气春生夏长、阴气秋收冬藏的变化。一日四时之中，天有白昼、黑夜的阴阳交替，人也有阴阳消长的不断变化。

  一般说来，早晨至中午，随着日头升起，人身阳气渐旺，阴气内守，精力充沛，工作效率高。中午至黄昏，随着日薄西山，人身阳气渐消，阴气渐长，虽仍有较强的活动能力，但逐渐感到疲倦。入夜之后，随着日没月出，人身的阳气逐步潜藏，阴气布于全身，需要合眼休眠。

在《孙子兵法》中，孙子曾以人体的这种变化来比喻军队在不同作战阶段的士气，提出了"朝气锐，昼气惰，暮气归"的著名原则。直到今天，"避其锐气，击其惰归"仍然是一条重要的作战原则。

有趣的是，人体阴阳盛衰的这种变化，与现代生物学家所得出的生物钟理论基本一致。生物钟理论认为，一天24小时之内，人的活动能力呈规律性的变化。2点至4点，身体大部分功能处于抑制状态，是通宵工作者效率最低的阶段；10点，注意力和记忆力最好,工作效率最高;13点至14点,体内激素变化,人感疲倦;15点,性格外向的青年人创造力旺盛,性格内向者则处于退潮时刻;22点,体内多种功能处于低潮,必须准备休息。

如果说,在兵书上,在军营里,军人的起居、着装、饮食、行走都有条令详细规定、严格管理的话,那么,医学家们对于人们的起居也有若干要求,二者所不同的是:条令对于军人是规定,医家对于人们是劝诫。

万变不离其宗,劝诫的内容还是"起居有常"四个字。不同的是把这"常"字讲得更深刻、更细微、更系统而已。

清朝曹廷栋指出:"衣食二端,乃养生切要"(《老老恒言·省心》)。衣食不在于追求锦衣玉食,只要做到"食取称意,衣取适体,即是养生之妙药。"具体地说,何谓"衣取适体"呢?诸如衣着要整洁,衣裳要勤洗勤换,宽紧要适当。甚至地域不同要求也不同。北方寒冷,因此"若要安乐,不脱不着";南方炎热,因此"若要安乐,频脱频着"。诸如此类,都是穿衣戴帽上的"起居有常"。

又如住和行,古人要求"屋无高,高则阳盛而明多;屋无卑,则阴盛而暗多"(《天隐子·安处》);要行如风,坐如钟,立如松。

古代医家还十分重视睡眠,认为"养生之诀当以睡眠居先。睡能还精,睡能养气,睡能健脾益胃,睡能坚骨强筋",睡眠是"药",而且是"非疗一疗之药也,乃治百病、救万民、无试不验之神药也"(《笠翁一家言全集·闲情偶集》)。尤其是老年人,睡眠好,大有益。所以苏东坡有诗云:"此身正似蚕将老,更尽春光一再眠。"

在起居上,中国养生学认为春季宜早睡早起,常信步漫行于庭院,舒展肢体,与大自然融为一体,顺时养神;老年人中午适当安排时间午睡,缓解春困,消除疲劳。夏季,如《黄帝内经》所言:"夏三月,此谓蕃秀,天地气交,万物华实,夜卧早起,毋厌于日,使志无怒,使英华成秀,使气得泄,若所爱在外,此夏气之应,养长之道也。"秋季起居,宜早卧早起,早睡顺阴精的收敛,早起应阳气的舒展。冬季起居,宜早卧晚起,早卧利于蓄养人体阳气,晚起利于保护人体阴气,使人体阴阳平衡,以抵御自然界阴寒之气。

总之,关于衣食住行各个方面都要尊循生活规律,都要守"常"。古人箴言如醇,凡欲长寿者不可不遵行之。

## 13、利胜者辱,过养则病

中国的传统哲学有一个明显的特征就是讲求"中和",适可而止,不偏不倚。反对"过",认为过犹不及,物极必反。

这一思想方法反映在军事上,就表现为反对穷兵黩武。《吴子·图国》明确主张:"天下战国,五胜者祸,四胜者弊,三胜者霸,二胜者王,一胜者帝。"《孙膑兵法》也说过:"利胜者辱。"穷兵黩武,欲壑难填,胜仗打得愈多,亡国败军也就来临了。

古人从中受到启发,认识到养生健体也应当把握一个限度,如果养生之法,也象穷兵黩武那样,"谋得战胜,兼弱取乱,以致骄逸之败"(陶弘景《养生延命录》),那就适得其反了。明朝祝允明指出:"彩色所以养目,亦所以病目。声音所以养耳,亦所以病耳。耳目之视听所以养心,亦所以病心。中则养,过则病"(《读书笔记》)。"中"就是中和,无太过,无不及,恰到好处。

随着人们生活水平提高,电视机越来越普遍。日益丰富的电视节目给人们的业余时间增添了许多乐趣。但是,许多事实证明,无节制地长时间看电视,有损人体健康。电视机的光线亮度不均匀,人们看电视时,眼部肌肉处于紧张状态,眼和屏幕的距离不适当,或长时间目不转睛,眼睛肌肉的调节力就不得不增大,因此而感到疲劳不堪。眼肌疲劳之后就会觉得眼睑沉重、眼帘干涩,有灼热、异物感和压迫感,眼球发胀、痛疼等等,继而出现视物模糊的症状,直至引起视力减弱。所以,电视不可不看,也不可久看。

饮食营养是维持生命的最重要因素。平衡合理的营养对健康长寿来说是至关重要的。《黄帝内经》中说:"饮食有节,起居有常,可度百岁。"民谚也说："若要身体好,不饥不过饱。"事实证明,饮食过度,是导致疾病和早衰的关键因素之一。关于这一点,我们已在"兵非多益,食非过益"一节中详细阐述,在此不再重复。

音乐是人类的精神食粮,给人类生活增添了丰富的色彩。一曲威武雄壮的交响乐,能振奋人心,激励斗志。一曲悦耳动听的轻音乐,能使人心旷神怡,胸怀舒展。古人有"听弦歌,知雅意,通性灵"之说,可见音乐对人的精神确实很有好处,不仅可以陶冶人的情操,而且能调节人的精神生活,使人精神焕发,从而起到养生的作用。但是,目前有些人盲目追求音响刺激,长时间沉醉于强烈刺耳的爵士乐、摇滚曲之中,有的人则整天戴着耳塞欣赏音乐。物极必反。这种过度的音响刺激必然引起大脑皮层兴奋与抑制平衡的失调,导致条件反射异常而引起中枢神经系统、心血管系统、内分泌系统及消化系统疾病。如血压和血中胆固醇升高、心率不齐、视力减退、胃溃疡病等增多,对传染病抵抗力降低,学习和劳动效率下降。可见,持久的噪音会使人早衰。

总之,无论是视听,还是饮食劳逸,都应当适可而止,切勿过偏过极,如此才能真正收到养生的效果。《黄帝内经》中"生病起于过用"这句话,虽然流传二千多年,至今仍耐人寻味。

## 14、欲不可绝，欲不可纵

中国古代在性学研究上同西方各国有所不同，其重点既不在于生殖繁衍，也不在于性享受，而是为了养生长寿。于是，千百年来，佛、道两家为此分宗立派，争论不休。争论双方针锋相对，一派主张禁欲，一派主张纵欲。主张禁欲者认为，精液是生命之根，元气之本，不可耗费，否则有损寿命。所以有的医书伪托彭祖之口，声言"上士别床，中士别被，服药千裹，不如独卧"，主张禁止房事。而另一派则认为男为阳，女为阴，阴阳交合，可以取阴补阳，延年益寿。这就是所谓房中采补说。男可补女，女可补男，甚至认为"能一日而数十交(交媾)而不失精者，诸病皆愈，年寿日益"(《玉房指要》)。

在这场佛、道关于房中养生的争论中，兵学家也有站出来提出批评的。例如成书于汉代的兵书《阴符经》就旗帜鲜明地反对纵欲，认为："淫声美色，破骨之斧锯也。世之人若不能秉灵烛以照迷津，伏慧剑以割爱欲，则流浪生死之海，害生于恩也。"它希望人们用智慧之烛照亮迷茫之路，用智慧之剑割断情爱之欲。

在"纵欲"和"禁欲"两种倾向长期并存的同时，一些具有科学态度的医学家和思想家从深入研究人体的生理规律出发，逐步认识了性的生理特性，提出了一些有助于廓清人们认识误差的见解。大体说来，有以下几点：

一是欲不可无。儒圣孔子说过："饮食、男女，人之大欲存焉"(《礼记·礼运》)。素有儒家亚圣之称的孟子也明确指出："食、色，性也"(《孟子·告子上》)。他们认为男女性欲与饮食之欲是人类与生俱来的基本属性,基本需求。中国晋代医学家葛洪说:"人复不可都绝阴阳,阴阳不交,则坐致壅阏之病,故幽闲怨旷,多病而不寿也。"说明人断绝了性生活会多病短寿。

二是欲不可早。按照《周礼》,男子三十岁娶妻,女子二十岁出嫁。这当然是一个晚婚年龄。至于男婚女嫁的年龄阶段,那是以男子二十岁举行冠礼,女子十五岁允许出嫁作为标志的。

三是欲不可晚。古代医学家认识到,男女的性功能有一个自然的成长过程,早婚、早育固然不好,但是如果年事已高,"阴已痿而思色以降其精,则精不出而内败,沁便涩而为淋;精已耗而复竭之,则大便牵痛,愈痛则愈便,愈便则愈痛"(《妇人良方·精血篇论》)。意思是说,老年人阴茎已经萎软而仍旧强行交媾以博一时的泄精之乐,那么,就会精液射不出,身体内部严重受损,小便不畅演变成淋病。久而久之,大小便都会发生疼痛感。

四是欲不可纵。唐朝吕洞宾有一首《警世》诗这样写道:

"二八佳人体似酥,
腰间仗剑斩凡夫。
虽然不见人头落,
暗里教君骨髓枯!"

春秋时,晋平公就是这种骨髓枯的人。郑国的大夫子产出使晋国慰问晋平公时发表了一通意见,成为历史上养生学的要言妙论。他说:人们正常的工作生活程序应当是上午听政(听情况汇报),下午访问(下基层调查研究),晚上修令(提出改进措施),夜间休息。可是晋平公把全部时间都用到女色上面,这就必然生病。美人过多地集中在一人身上,又毫无节制地纵欲,生命就危险了。两千多年前子产能讲出这一番高论,自然是弥足珍贵的。明朝有一个叫赵三翁的养生家索性一语破的:(性欲是)"生尔处乃杀尔处"(《昨非庵纂·颐真》)。这一警世之言,倒也颇有哲理意味。

对于性欲,正确的态度是掌握规律,适应规律。《黄帝内经》认为,人的性欲与生殖是人在生、长、壮、老、死过程中的生理本能。人体性机能的盛衰和生殖能力的变化,主要取决于肾气。肾气有一个由弱到强,由盛到衰的过程,由此决定了人体性机能和生殖能力的发展曲线。一般女子14岁、男子16岁左右,肾气已经充盈,出现月经按时而行和精液盈满外泄的现象,标志着性功能成熟,若两性交配便能生育子女。男女40岁左右,肾气开始虚亏,性功能逐渐衰退。女子49岁、男子64岁左右,肾气将近衰竭,出现月经终断和精液稀少的状况,标志着性机能的衰弱和生育能力的丧失。由于肾是五脏六腑之精的收藏之所,肾气的盛衰也同时影响到全身机能的运转,决定着人的生长、发育、健康与衰老。所以,中医对性生活总的观点是:毋需禁欲,亦不可纵欲,"惟有得其节宣之和,可以不损"(《抱朴子》)。

所谓"节宣之和",意思是说性生活应根据年龄和身体情况而适当节制和宣泄。孙思邈认为在正常生理情况下,"人年二十者(20～29),四日一泄;三十者(30～39),八日一泄;四十者(40～49),十六日一泄;五十者(50～59),二十日一泄;六十者(60～69),闭精勿泄,若体力犹壮者,一月一泄"(《千金要方》)。现代医学的观点与此大体一致。只是不同的夫妇存在体质差异,没有一个固定的性交次数模式。新婚夫妇,性的要求迫切,性交次数频繁,婚后数月,性交才能逐渐减少。有的夫妇,性交频繁,长久不衰,就不能以次数计算了,应当看性生活的第二天,是精神愉快,还是头重脚轻,精神疲倦。一般年轻的夫妇每周1～2次为宜,中年以上的夫妇可每周一次左右。老年人随着性欲和性功能的减退,可半月一次,或一月一次,但是有的健康老人,也可和年轻人一样。中国学者曾对260名长寿老人抽样调查,表明性生活可以调节老年夫妇的情绪,从而使他们晚年生活充实,增加生活的自信,利于健康长寿。调查中,二对百岁夫妇,80岁时还能进行房事,另一对长寿老人,90岁时还像年轻夫妇一样,每周过一次性生活。老年人完全禁欲、绝房事,有损于健康。老年人如果长期避免性生活,睾丸或卵巢、脑下垂体前叶的促性腺功能都会下降,性激素分泌减少,从而加速衰老过程,并容易导致生殖器官废用性、退行性变化。相反,适当的性生活不但可以延缓生殖器官废用性萎缩,而且可以延缓大脑衰老,促进老年人身心健康。

"节宣之和"还有另外一层意思,那就是夫妻进行房事时要讲究适当的方法,力求快慢、轻重皆相宜。孙思邈指出,房中交接应以"男女感动"为前提,"凡御女之道,不欲令气未感动、阳气微弱即以交合。必须先徐徐调和,使神和意感,良久乃可令得阴气。"马王堆出土的西汉竹简《合阴阳》则具体记述了涉及性交心理和方法的"戏道"。它把性交分为五个步骤,"一曰气上面热,徐吻;二曰乳坚鼻汗,徐抱;三曰舌薄而滑,徐屯;四曰下汐股湿,徐操;五曰嗌干咽唾,徐撼"。一个"戏"字,说明这五个步骤的实施过程中夫妻双方始终应保持游戏般愉快的心情,先有性感上的和谐和爱抚,使两情逐渐融合,然后慢慢地由外向内渐次深入,逐步达到性高潮。

现代医学研究表明,能否和谐交合,与掌握男女性功能特点关系密切。一般来说,男人性冲动来得猛烈,进入快感高潮迅速,从性交开始到射精,大约只需 $2\sim6$ 分钟,但性欲消退也比较快。女人性冲动相对较慢一些,一般要 $10\sim30$ 分钟才能达到快感高潮,但高潮维持时间长,性欲消退也比较慢。另外,男人性感点多集中于性器官,性交欲望强烈;女人性感点分布广泛,除性器官外,胸、背、颈、腿等处都有较强的性感,通过多方爱抚之后,才有性交的要求。基于这种明显的差异,性交时,男方应当有所"节",自觉控制性冲动,耐心地、温柔地爱抚妻子,以激发其性兴奋,女方则要有所"宣",应表现出积极的意愿,激动的感情,主动配合男方。交合之后,男方不要马上独自入睡,还应继续给女方以温存、爱抚,使彼此都感到酣畅、完善。

房室养生,不仅要讲究方法,而且还要注意性保健和性卫生,即古人所说的"入房有禁"。历代医家认为很多疾病都与房事不慎有密切关系,于是提出了许多禁忌。其中最主要的是:

①酒醉不宜同房;
②过度劳累之后不宜同房;
③忿怒之中不宜行房;
④妇女经期、妊娠期、产褥期、哺乳期不宜房事;
⑤患病期间要慎行交合;
⑥气候剧变时不宜交媾;
⑦夜半之时勿作云雨;
⑧远途归来不宜同房。

性生活不仅仅是夫妻性器官的交合,还需要心理上的满足,男子的性欲可以通过性器官得到满足,而女子的性欲更需要在语言和情感交流中得到满足,这是男性应该注意的。

## 15、治军要激励士气,养生要勤练气功

在中国传统的养生学中,吐呐导引之术占有极其显著的重要地位。几千年来,这枝放射异彩的奇葩始终显示出它的神奇功效,成为人们强身健体、延年益寿最为喜好的运动形式。

"吐故纳新"—吐出体内混浊的废气,吸收新鲜的空气,这种导引之术就是健体长寿的气功。气功始于何时,确切年代无从考究,但早在殷商的甲骨文中就有"气"字。气在人体中运行,断了气,人的生命也就终止了。由于"气"是无形的,无时不在,无处不在,于是逐渐成为一种抽象的概念,成为万物的本原。

春秋时代,哲学家、思想家们为了反对鬼神论,坚持唯物论,于是把"气"看作是一种精神现象。精神旺盛称之为"生气",情绪低落称之为"死气"。《孙子兵法》正是在这个意义上使用了"气"这个概念。军队斗志昂扬,称为"锐气",军队精神颓靡,称为"惰气"、"归气"。因此,孙子主张,未战之前,应当"谨养而勿劳,并气积力",搞好休整,蓄养和保持旺盛的士气;既战之时,要激扬士气,他说"故善用兵者,避其锐气,击其惰归,此治气者也。"总之,在《孙子兵法》中,关于"气"的论述都是指的军队的精神状态。

《孙子兵法》关于"气"的论述与中医理论的"气"是不是一致呢?

要回答这个问题,必须知道在春秋战国时代,思想家、哲学家们已把"气"看作了万物化生的动力。比如,孔子关于"血气"的论述就是在这个意义上使用的。他说:"君子有三戒:少之时血气未定,戒之在色;及其壮也,血气方刚,戒之在斗;及其老也,血气既

衰,戒之在得"(《论语》)。"血气"在这里既是生理的,又是心理的;既是物质的,又是精神的。"血气"在军事上就表现为"士气"。

从人的生理角度来看"血气",就应做到在五脏、六腑保持和培植"血气"。那么,如何培植呢?《韩非子·解老》说:"民少欲则血气活而举动理。""少欲"就要戒色、戒斗、戒得。

"色"、"斗"、"得"等等都是邪气。"邪气袭内,正色乃衰"(《管子·形势》)。为了扶正祛邪,就应"吐故纳新",就应加强气功锻炼。

气功的养练方法,自古以来多种多样,南朝陶弘景关于气功的"六字气诀"说得较为简明易学,摘引如下:

"常以鼻引气,口中吐气,当令气声逐字吹、呼、嘘、呵、嘻、呬吐之。若患者依此法,皆须恭敬用心为之,无有不差,愈病长生要术"(《养性延生录》)。

陶弘景的"六字气诀"就是气功,就是气功养生之术。

我们从中得到一个重要的认识就是,气功并不神秘,基本规律就是调息入静,这样就可以防治气衰神竭。你看他这"六字气诀"说得多么简明!这就是古人讲的"真传一张纸,假传万卷书。"

现在有些伪气功师,不仅把气功说得神乎其神,天花乱坠,更危险的是把气功引入了唯心论和神秘论,附会了迷信色彩。我们应当坚信,气无吉凶,气也不能干涉人事和社会生活。如果以"气"解释社会问题,预测人的吉凶祸福,那是毫无意义而且反科学的,应予摒弃。

## 16、战必选将，疾当择医

"置将不善，一败涂地"(《史记·高祖本纪》)，这是中国历史上以善于任将而著名于世的汉高祖刘邦的名言。一千多年后，明太祖朱元璋说得更明确："将必择有识、有谋、有仁、有勇者"(《明太祖宝训》)。打仗要选将，是因为战场之上有良将，也有庸将。秦赵长平之战中，因为赵王选将不当，任用了只会纸上谈兵的赵括为帅，赵军惨败长平，险些亡国。又比如，三国时被誉为智慧化身的诸葛亮，也因为选将不当，任用"言过其实"的马谡，招致街亭失守，使得蜀军很有希望取胜的攻魏战争无功而还。

选将之所以重要，因为实际生活中有良将与庸将之别。同此道理，择医之所以重要，也因为现实生活中有良医与庸医的存在。对于择医治病，古人感慨甚多。明朝顾炎武说得好："古之时，庸医杀人。今之时，庸医不杀人，亦不活人。使人在不死不活之间，其病日深而卒至于死"(《日知录》)。现在，医学虽然已很发达，但是浪荡里巷骗人钱财的江湖医生也不少见。

例如，有一患者，初秋之时，忽然觉得颤栗恶寒，身体发热，并伴有咳嗽，找一江湖医生治疗。江湖医生用表散之药治疗后，不但没有疗效，身体还日渐瘦弱。进入冬天后，又加服人参、白术补剂，转觉昏昏欲睡，不思饮食，浑身乏力，泻痢不止，生命处于垂危状态。患者亲属只好将其送往医院，经医生及时抢救和治疗，患者才脱离危险，逐渐恢复健康。此案中的江湖医生，见病人有寒热，即疑其为外感，随意用发散之药治疗。在病人久治不愈的情况下，又疑其为虚损，而采用补益之药。其整个治疗过程中，既不究病因，又不辨病情，几乎将患者活活治死。

遇上这种庸医，倒不如不治。常言说，"有病不治，常得中医（中等水平的医治）"，因为不少疾病可以依靠自身的抵抗力慢慢康复。这虽然给自己带来一些病痛的煎熬，但总比碰上庸医，把小病弄成大病，甚至一命呜呼要强得多。所以，《折肱漫录》说："欲求速效，而轻用医药，为病者所忌。药者，人生之大利大害也。不遇良医，不如不药。不药而误也悔，药而误也亦悔，然不药之悔小，误药之悔大。"当然，这不是说不及时求医早治，而是说比找江湖医生要好得多。

江湖医生古今都有，君不见某处墙角或电线杆上那些"祖传秘方"、"专治艾滋病"花花绿绿的招贴，皆是江湖派所为。一名八岁癫痫病患者，吃了那里的"秘方"，癫痫病未见效果，反而增加剧烈腹痛。病家只得改弦更张，到医院请专家诊治。结果发现，小患者因服"秘方"，已受到严重铅中毒损害。这正像《黄帝内经》所批评的："粗工凶凶，以为可攻，故病未已，新病复起。"江湖医生危害虽然不小，但其人数毕竟有限，而作为职业医生，则应该对技术精益求精，以防发生误诊、误治。

例如本作者曾收到一份从湖北省某地寄来的病历，一位19岁的姑娘因为心烦失眠、急燥易怒而不能学习和工作，在其当地一直以"青春期精神分裂症"诊断和治疗而效果不好。经过仔细察看这份病例的全部材料，诊断为"癫痫病、精神运动型发作"，遂给予相应中药治疗，五个月后症状显著缓解，而能参加正常学习。

对于患者,固然不能病急乱投医。对于医生,则应当继承古代重视医德医术的优良传统。清代医学家吴塘告诫道:"生民何辜,不死于病而死于医,是有医不若无医也。学医不精,不若不学医也"(《温病条辨序》)。

## 17、官兵一体,医患同心

将士一体,官兵团结,同心同德是取得战争胜利的重要条件之一。要想做到将士一体,除了"铁的纪律"之外,最重要的则是将领应在士兵心目中具有崇高的威信。而威信的树立,关键在于将领的战绩与行为,凡深谋远虑、指挥得当、百战百胜的将领,必然有极高的威信;善于严格训练又能爱护士兵,并以身作则的将领,更能使他的军队团结一心,士兵乐意服从指挥甚至为之效命。

医患关系虽然与官兵关系不同,但是作为真正高明的医生,要想提高治疗效果,要想使医学知识得到传播和推广,同样必须具备关心爱护病人,严格要求病人,并且严于律己以身作则的品格。孙思邈在其所著《备急千金要方》的一开头便提出:医生必须爱护病人,急病人之所急,待患者如亲人。为人治病应该不避艰辛,不计报酬,全力以赴。他说:"凡大医治病,必当安神定志,无欲无求……若有疾厄来求救者,不得问其贵贱贫富,长幼妍蚩,怨亲善友,华夷愚智,普同一等,皆如至亲之想。亦不得瞻前顾后,自虑吉凶,护惜身命。见彼苦恼,若己有之,深心凄怆。勿避崄巇,昼夜寒暑,饥渴疲劳,一心赴救。"视一切病人"皆如至亲",不带一丝私欲杂念,不避艰险,不惜自家身命,一心救护病人,是对医生的基本要求之一,也是医生在技术上精益求精、不断提高疗效的前提条件。

除了关心、体贴病人之外,医生还应从疾病的需要出发,给病人以一定的约束,如饮食忌口、劳逸的度与量、情志的喜怒等,均应提出明确而严格的要求,否则也不能收到满意的治疗效果,或者使已经取得的效果前功尽弃。《黄帝内经》即曾指出医生必须避免

"五过",其中第四种过错就是"医不能严,不能动神,外为柔弱,乱至失常,病不能移,则医事不行,此治之四过。"指出医生不敢于严格要求病人,表现出软弱无能,以致病人不遵医嘱,而导致气血紊乱,疾病不能痊愈。医生虽然不能像将军那样订立若干惩罚条款,但也必须有使病人"动神"的措施,以保证"医事"活动正常进行。

宣传、教育,指导民众保持良好的卫生习惯,注意锻炼身体和精神修养,以预防疾病发生,同样是医生的一个重要职责。但是,医生不仅是一个教育者,而且更应该是民众的典范。医生只有自己身体力行,以身作则,才能够使其教育活动具有说服力。否则,再高明的医学理论、正确的养生防病技术,也难免被视为空洞的说教而失去意义。《黄帝内经》记载推广养生保健知识时提到"上古圣人之教也,下皆为之。"唐代初年医学家杨上善对这句话做了正确的解释,他说:"上古圣人使人行者,身先行之,为不言之教。不言之教胜有言之教,故下百姓仿行者众,故曰下皆为之。"圣人,指道德高尚的人。下,指民众。身教胜于言教,圣人的行为是民众的楷模,所以百姓自然听从其言而仿效其所作所为。

在医疗工作中,经常遇到患者不遵"医嘱"而使疾病加重或复发的事。例如服中药时常有"忌口"的要求,也就是医生根据病情和所用药物的特点,对病人的饮食加以限制,否则会影响治疗效果。一位患"精神分裂症"的青年男子,服中药后痊愈,不仅能参加正常工作和学习,而且结婚得子。医生曾告诉该男子:"病愈后也要忌酒",但时过几年,却因耐不过亲朋劝让,多饮几杯别人的"喜酒",次日疾病复发,重新住进医院。又经过较长时间治疗,虽病愈出院,却未能避免妻离子散的悲剧。又如一少年癫痫病患者,服中药大有好转,但因到亲戚家误犯"忌狗肉"的医嘱,以致食后一小时便癫痫大发作,使身体遭受不应有的损伤。

## 18、相敌以达变,四诊以察因

相敌,是《孙子兵法》中的一个术语,意思是侦察敌情。在《行军篇》中孙子列举了三十二种"相敌"之法,诸如"敌人使者措词谦逊却又在加紧战备的,是准备进攻;措词诡诈而强硬做出前进姿态折,是准备后退;轻车先出动,部署在翼侧的,是在布列阵势;尚未受挫而来讲和的,是另有阴谋;兵卒奔走而布列兵车的,是期待与我决战;杖而立者,饥也;汲而先饮者,渴也;见利而不进者,劳也"等等。这些方法有的侧重于由此知彼、由表知里,有的侧重于由近知远、由小见大。其基本原则是要求透过现象看本质,从种种表面征候辩证地分析敌情的变化。

先秦时期,河曲之战中晋军坐失良机的教训就充分说明了这一点。公元前615年,秦国攻打晋国,晋王派赵盾率军到河曲(今山西永济)迎战。晋军针对秦兵出国远征、难以持久的弱点,采取深垒固军、待其撤退而击之的策略。秦军因久战不胜,决定撤退。为掩饰其意图,秦军将领派使者以强硬的言辞约晋军于第二天决战。晋军的一位副将从秦使的眼神和语气中察觉到秦军要撤退,建议乘机予以截击。可惜赵盾不懂得"辞强而进驱者,退也"的相敌之术,未采纳这一建议,致使秦军在当晚得以安然撤走。孙子的这些相敌之法,尤其是其中蕴含的思想方法对后世影响较大,除兵家将之奉若法宝之外,向来认为治病如驱敌的医家也对此颇为重视。明代医学家孙一奎曾在《医旨绪馀》中说:"按图用兵而不达变者,以卒与敌;执方治病而不察因者,未有能生人者也。"这里的"卒"与"猝"相通。全句的意思是说,用兵打仗如果只知机械地搬用兵法,而不能及时掌握敌情的变化并加以灵活处置,必然常常处于仓促应战的被动地位;用药治病如果只知刻板地套用药方,而不善于了解疾病的原因并适当处置,则很难达到治病救人的目的。

无论是顺手掂来,还是深思熟虑,以用兵达变喻治病察因之语,实在是神来之笔。神就神在它一语切中了兵医相通的关键之一。

与孙子重宏观、整体、辩证的思维方式一样,历代中医也将人体视为一个整体。他们认为人体的各个部分通过经络保持着密切的、不可分割的联系。人体内部的病变,必然反映到外部,表现为面色、精神、舌象、脉象等多方面的异常改变。

  例如有病而面色赤者多为心脏热盛,有病而面色苍白者多为肺脏不足,面色萎黄者多是脾脏虚弱,面色青者多是肝脏病或痛证,面色黧黑者多是肾脏病。精神爽慧是健康的表现,急燥易怒多因肝气有余,无故悲哭多属肺气不足,嘻笑不休是因为心脏热盛,不分场合"总想唱歌"其病在脾。在脉象方面,中医诊病通常把脉象分为28种,每种脉象反映不同的病情。以五脏病概括言之,肝脏病的脉象如琴弦,肺脏病的脉象轻浮"如水漂木",脾脏病的脉象软弱无力,心脏病的脉象节律不调,肾脏病的脉象沉小而坚硬。

所以,中医诊断疾病的方法与西医明显不同,一般不需借助仪器和化验,仅凭医生的感官来获取临床资料,然后通过逻辑思维,加以分析、综合、推理,即可作出诊断。元代名医朱丹溪揭示了其中的奥妙。他在《格致馀论》中说得十分清楚:"欲知其内者,当以观乎外;诊于外者,斯以知其内;盖有诸内者,必形诸外。"这也就是中医诊断学的基本原理。

基于这种原理,历代中医在长期的实践中逐步摸索出四大类有效的诊断方法,即望、闻、问、切。

**望诊**,主要观察病人的神色、形态、舌苔、大小便和其他排泄物等,如形体肥胖之人多痰湿,尤其要注意预防中风半身不遂;形体削瘦者多火热,如果不是胃肠病应该检查一下血糖,排除糖尿病;病人弯腰抱腹来就诊,多是腹痛;双肩高耸者,多是久病咳喘;腰部动转不利而痛者,多是肾脏不足之病。察看病人神色具有重要诊断意义,尤其是眼神更应注意观察,俗话说"眼为心灵之窗",中医理论说"五脏六腑之精华皆上注于目。"病人呈惊恐之状,多有剧痛在身;神色木呆,多是精神性疾病,或长期服用了镇静剂;危重病人精神萎靡,忽然精神焕发,两眼异常光亮,应谨防突变,恐为"回光返照",命在须臾;就医者脉象无异常,眼神飘忽不定,说话时时咽唾,可能有伪并非真正是病。

**闻诊**,主要听病人的语言、呼吸、呻吟、喘息、咳嗽等声音的高低、强弱。例如话语过多,喋喋不休,或沉默寡语,中医诊为肝气不舒,西医可诊为神经症;前言不达后语,答非所问,或骂詈不避亲疏者,属于心神紊乱,病名为"狂",西医则名"精神分裂症";语声低怯不能接续者,是十分虚弱的表现。嗅察病人的口气、痰涕和大小便的气味等,如排泄物粘浊而且气味重者,多属热症;而排泄物清稀气味小者,多为寒症。《黄帝内经》概括得好,谓:"诸病水液,澄彻清冷,皆属于寒";"水液混浊,皆属于热"。这里的热和寒,是对疾病两种不同性质的高度概括,根据疾病性质,中医便可以开出相应的药方来。例如尿色黄,排尿时有痛感,应是热证,可以服用导赤丹治疗。尿色清,而且频数并无痛感,是寒证,可以用金匮肾气丸来治疗。

**问诊**,主要询问病人的寒热、头身、二便、饮食、胸腹、病人的既往病史、发病经过、现在症状、生活习惯及治疗服药后的情况等。

**切诊**,包括脉诊和触诊,主要是按病人的脉象和触摸肌肤、胸腹、胁肋等部位有无异常情况。例如触摸皮肤可以了解病人的体温和皮肤的干涩或润滑情况;触按胁肋、腹部,可以了解有无压痛及疼痛的部位和程度;检查腹部可以了解有无包块及其形状、软硬和移动性;按压面部和下肢,可以检查患者有无浮肿及其程度等,都对诊断病疾有直接参考价值。

疾病的发生发展是一个十分复杂的过程,想早期正确地认识它,单靠某一种诊断方法显然是不够的。所以中医学在论述诊法时,既强调精于四诊,又主张四诊合参。

首先,作为一个医生如果不能精通四诊之法,就不能算合格的医生。《黄帝内经》对四诊的基本功提出了很高的要求,即:"见其色,知其病,命曰明;按其脉,知其病,命曰神;问其病,知其处,命曰工。"其中的"明"、"神"、"工"为医生诊法达到上乘的标志。《难经》发展了这一思想,指出:"望而知之谓之神,闻而知之谓之圣,问而知之谓之工,切而知之谓之巧。"其核心思想是要求医家深入研究四诊之法,达到"神"、"圣"、"工"、"巧"的最高境界。至于怎样才算达到这一境界,书中进一步解释说:"望而知之者,望见其五色(青、赤、黄、白、黑)以知其病;闻而知之者,闻其五音(角、徵、宫、商、羽)

以别其病；问而知之者，问其所欲五味，以知其病所起在也；切脉而知之者，诊其寸口，视其虚实，以知其病在何脏腑也。"

其次，四诊之法虽然各具特色，但同时也各自存在着这样或那样的片面性，只有把它们综合起来使用，才能够扬长避短，所以中医学特别强调"四诊合参"。所谓"四诊合参"，主要有两层含义。一是要求综合运用望、闻、问、切四种诊法，不可偏其一而废其三。在通常的情况下，容易只注意脉诊而忽视望、闻、问诊，因而很难全面而深入地掌握病人的病症及病因，往往得出的是片面的认识，这是应当力戒的。二是要求将四诊所得的有关病史、症状、形态、神色和脉象等材料，互相进行综合比较，经过去粗取精、去伪存真的分析，辨别疾病的性质与病机变化，从而作出确切的诊断。尤其是脉象与某些症状表现不一致时，更应详加辨别，在最后诊断上或"舍脉从症"，或"舍症从脉"。

## 19、运筹以决胜,辨证而施治

清朝时候,《孙子兵法》研究又是一个高峰期,著书立说的学者多达一百三十余人,其中有一个叫邓廷罗的学者声名赫赫,著有《孙子集注》、《兵镜备考》、《兵镜或问》。他在《兵镜备考》中说过的一段话颇有份量,引起时人的重视。他说:"救乱如救病,用兵如用药。善医者因症立方,善兵者因敌设法。孙子十三篇,治病之方也。"这里有一个现象值得深思。为什么兵学家讲"用兵如用药",医学家讲"用药如用兵","兵"与"药"可以相互借鉴吗?可以。

原来,这里有一个十分深刻的道理。
我们知道,中医不同于西医。西医注重细微的具体的研究,如对病毒、细菌等等进行深入的、定量的分解和分析。中医则注重宏观整体的研究。也就是说它对疾病产生的原因进行多方考察。《黄帝内经》说:"夫百病之所生者,必起于燥湿寒暑风雨,阴阳喜怒,饮食起居。"汉代张仲景在《金匮要略》中指出,疾病发生主要有三条途径,即:"千般疢难,不越三条。一者,经络受邪入脏腑,为内所因也;二者,四肢九窍,血脉相传,壅塞不通,为外皮肤所中也;三者,房室、金刃、虫兽所伤。以此详之,病由都尽。"晋代陶弘景在《肘后百一方·三因论》中提出了"三因说",即"一为内疾,二为外发,三为它犯。"总之,中医讲究的是五运六气、四诊八纲、脏腑经络。一言以蔽之,辨证施治。

中医的辨证施治与中国古代兵法的运筹决胜,从哲学上讲都是系统分析的思维方法，都是重宏观、重整体、重综合的思维方法，所以中医之用药与中国古代兵家之用兵二者之间可以沟通，可以联接，可以相互为用。

以《孙子兵法》为例，它的系统思想相当丰富，它提出的"经五事"、"校七计"就是从战争总体上对双方政治、经济、军事等基本条件进行考察，揭示其内在的关系，因此，它的决策方法是系统分析的方法，坚持整体性和优选性原则。

中医辨证施治正与古代兵法这种系统分析的运筹决胜思想相同，它治病的特点在于不以手术、注射为主，而以内服中草药为主，往往只需对症下药，让患者服用几付汤药，便可以药到病除，起死回生。

辨证施治有它自己完备而系统的机理。所谓"辨证"，就是对望、闻、问、切四诊所得的有关病人的疾病信息进行综合分析，辨别出疾病的原因、性质、部位，抓住疾病的本质，然后进行论治。在这个复杂的思维过程中，关键是概括出反映疾病性质的"证"。

证，是证候的简称，其含义是证据或征象。中医诊断学中的证候，不是一般的单个症状或某些综合证候群，而是综合分析各种症状和体征后，对于疾病处于一定阶段的病因、病位、病变性质以及正邪关系等情况的病理概括。可以说，"症"是辨别"证候"的基础，"证候"是各个"症"的综合。

诊断就是从诊察各种病情、病因开始，进而分析判断其内在联系，最终辨识证候的过程。诊与断紧密相连，缺一不可。如果有诊无断，则无异于军事上只去侦察敌情，不去分析和判断敌情。综观中国古代战争史，聪明的军事家从来都是二者兼重的。

以中医比较重视的八纲辨证为例，它有一个内容广泛、系统复杂的机理。"八纲"，即指阴、阳、表、里、寒、热、虚、实八类证候。任何疾病都可以用八纲辨证来分析归纳。如论疾病的类别，不属于阴，便属于阳；论病位的浅深，不在表就在里；论疾病的性质，不属于寒便属于热；论邪正的盛衰，不是正虚，便是邪实。下面这个具体例子更能说明中医诊断的思维特点：

明代有一个名叫卢廉夫的大夫，外出归来后感到劳累疲倦，体温较高，身体略痛而头不痛，自以为外感而用九味羌活汤治疗。服用三次之后，汗虽发出，但高烧不退。于是又连续服小柴胡汤五、六剂，体温反而更高。不得已，他只好请来另一位名为虞抟的大夫帮助诊治。虞大夫来到病床前时，注意到床边放着一杯已煎好的汤药，一问方知是大承气汤。虞大夫先诊其脉，感觉脉象不属于伤风感冒的征候，惊诧地说："你差点儿自杀了！你的病属内伤虚证，如果将此药一饮而下则必死无疑。"

卢氏不以为然地说："我平生元气充实，从来没有虚损证，明明是外感所致。"

"如果作外感病治疗，你的脉搏既不沉实，又没有头疼鼻干、潮热谵语等症状。如今治疗已有八天了，不应当仍然体热不退。"

卢氏无言以对，默认了虞大夫的诊断。于是虞大夫以补中益气汤加附子三分，请卢氏当天晚上连服二剂。第二天早晨，卢氏觉得病情没有什么好转，打算继续服用降温退烧的药。虞大夫坚持说：

"你再按我的药方吃两剂,如果没有效果便拿我问罪。"

果然,两剂汤药服下不久,体温降了下来。随后,虞大夫减去药中的附子,让卢氏继续服用二十剂,以逐渐强健元气。病愈之后,卢氏渐愧地说:"我几乎亲手杀死自己了。"

这个病案中,卢大夫仅凭表面症状,简单诊断,错误施治,险些害了自己的生命。虞大夫则与之相反,首先通过四诊以收集资料,辨别是外感还是内伤;其次是在鉴别诊断中注意抓疾病的主要临床表现或特殊体征,得出"内伤虚热"的判断;再次是在对症下药之后还在治疗中继续验证了自己的诊断。由此可见,中医诊断最忌讳头痛医头,脚痛医脚,简单从事,而高度重视诊、断结合,透过现象看本质,辨证施治。

## 20、先伐谋后伐兵，先食疗后药疗

"恨病吃药"，这大概是中外通用的俗谚。一则，大凡药物都有一定的副作用。如果不是"恨病"，何苦来惹上这个"副作用"，或又种下新的病根。二则，"良药苦口"，这大概以中药为甚。如果不是"恨病"，谁又愿意捏着鼻子喝下那苦涩难咽的药汤呢？

"恨病"能不能不吃药而采取别的途径也能治病呢？

唐代的大医学家孙思邈从军事上用兵作战的道理中回答了这个问题。他在《千金要方》中写了一段异常精辟的论述："夫为医者，当须先洞晓病源，知其所犯，以食治之，食疗不愈，然后命药。药性刚烈，犹若御兵。兵之猛暴，岂容妄发。发用乖宜，损伤处众。药之投疾，殃滥亦然。"在他看来，用药物去治病，如同用军队去打仗，虽然消灭了敌人，但造成的损失也是巨大的。用药治病，虽然治愈了疾病，但是药物给人体脏器造成的直接或间接的损失也是不小的。因此，他认为用药治病如同用兵作战，都是不得已而用的下策。那么，上策是什么呢？他的回答是"以食治之，食疗不愈，然后用药"。值得注意的是，孙思邈还明确指出过"能用食平疴，释情遣疾"的人，才能称作高明的医生。由此可见，孙思邈通过以药喻兵来阐述其"先食后药"的思想，是与"先礼后兵"，"先伐谋后伐兵"等兵家思想一脉相通的。清代名医徐大椿继承这一思想，在《医学源流论》一书中进一步明确指出："兵之设也以除暴，不得已而后兴；药之设也以攻疾，亦不得已而后用，其道同也。"

利用食物直接治疗某些疾病，确实如同通过伐谋、伐交手段和平解决某些政治问题一样，没有副作用，没有后遗症。对于食疗的认识,并不始于孙思邈,也不始于《黄帝内经》,在人类历史上,我们的祖先早在公元前21世纪的夏禹时代就触及到了食物的治疗作用。那个时候,人们就发现酒能舒经活血,可以治病。繁体字的"醫"字,之所以取了"酒"字的大半,就是因为"酒所以治病"(《说文解字》)。酒能治病,茶能治病,萝卜大葱等等都可入药。

《癸辛杂识》记载的一个故事是颇有意味的。说是宋朝有一位高官久咳不愈,病情已到了十分严重的地步。当地官府急忙四处寻找医生前往医治,但是无人敢于奉命。下面的官府为了应付差使,找了一位年过七十的乡村老医生。这位老医生自己也得咳嗽病,每天从早咳到晚。他在赴命途中又咳又渴,便向一户村民要点水喝。这户村民倒了一碗热水给他喝。他喝过之后,忽然觉得咳嗽好了许多,于是又要了一碗,咳嗽又好了许多。老医生惊奇地问:"这是什么水?"村民说:"穷乡僻壤,没有茶水,舀了一碗萝卜水给你喝。"老医生十分感激,并向村民又要了一口袋萝卜干。老医生在途中吃了几天萝卜干,咳嗽全好了。进入官府后,老医生经过诊断确认高官的病与自己的一样,便写了一副药方,并煞有介事地声明这药要由他亲自煎。煎时,他悄悄放进一些萝卜干。几天之后,高官果然病愈。高官高兴之余,一下就赏给老医生千余两黄金。原来,萝卜有祛痰止咳、去积导肺、宽胸利膈等疗效。

因为食疗具有这种"兵不顿而利可全"的效果,所以古代医家无不重视食疗,并将之与食养放在同等重要的位置上。《黄帝内经》中明确指出:"是故谨和五味,骨正筋柔,气血以流,……长有天命"。此后,"谨和五味"被后世医学家视为食疗、食养的基本原则。

五味,即食物所含的辛、酸、甘、苦、咸五种味道。五味与五脏的关系极为密切。《黄帝内经·素问》中说:"五味入胃各归其所喜,故酸先入肝,苦先入心,甘先入脾,辛先入肺,咸先入肾。"《黄帝内经·灵枢》则指出:"肝病禁辛,心病禁咸,脾病禁酸,肾病禁甘,肺病禁苦。"由此可知,五味与五脏之间具有亲和性与排斥性,因而各有所宜也各有所忌。宜则五味调和,五脏得益,气血通畅,形体壮实,人体健康;不宜则精气不足,疾病丛生,形体受损。既然食物的五味与五脏关系如此密切,并进而影响人体的气血津液和精气神状态,那么"谨和五味"无疑是调养身心,医治疾病的重要方法。

"谨和五味"主要包含两方面意思。一是避免偏嗜,二是适时调味。

偏嗜往往使某一脏腑之气过盛,五脏因失去平衡而受损,从而导致疾病。如长期摄入脂肪过多,超过机体需要,会诱发高血脂症、动脉粥样硬化及冠心病。而长期缺乏植物纤维,则会影响人体胃肠的代谢功能,产生一系列不良后果。因此,《黄帝内经》要求人们,平时饮食中要注意"五谷为养,五果为助,五畜为益,五菜为充,气味合而服之,以补益精气。""五谷"泛指粳米、大豆、小豆、麦、黄黍等谷类食物。"五果"泛指桃、李、杏、栗、枣等瓜果类食物。"五畜"泛指家畜、家禽等肉类食物。"五菜"泛指植物蔬菜类食物。如果将这些食物按照主次搭配,调和而食,必然对人体的精气有直接的补益。

调和五味不是一时之计,一年四季都不能忽视。因为四季气候的变化对人体五脏及精气神有着直接或间接的影响,而适时调整五味比重,增加或节制某些食物,则有利于保持体内五脏平衡,气血通畅。如冬季人们喜欢吃辛辣的调味品以促进食欲和刺激体温而感和暖,夏季则喜欢吃些阴凉消暑之品,使人感到清新适口。所以,古人对四季饮食有着不同的要求。

中医理论上自古就有"医食同源"的说法。这说明食疗与药疗之间并没有严格的界限,强调食疗并不意味着取消药疗。实际上,二者常常是同时配合,相辅相成的。一般情况下应以食疗为主,药疗为辅,但病情严重时,当然还是应以药疗为主,食疗为辅。

## 21、胜敌在得法，治病如治寇

"治病如治寇"，是明代医学家张景岳的一句名言。他精通医学，对兵学也颇有研究，因而常常以兵家的思维方式分析论证治病用药的方法。他的一个基本观点是，"治病如治寇，知寇之所在，精兵攻之，兵不血刃矣。"

自古知兵非好战。"兵不血刃"是古代兵家追求的最高目标，孙子最著名的观点就是"百战百胜，非善之善者也；不战而屈人之兵，善之善者也"。张景岳以此比喻治病，实在是妙不可言。它既明确了中医治病的最高标准，又道出了中医治病的显著特点。

与西医相比，中医治病的特点在于不以手术、注射为主，而以内服中草药为主，往往只需对症下药，让患者服用几付汤药，便可以药到病除，起死回生。《三国志·华佗传》记载了素有神医之称的华佗的行医事迹，从中可以明显地看出中医治病的特点。

有一天，官府小吏倪寻和李延同时患病，卧床不起，其症状都是头痛身热。华佗前往治疗时，首先仔细诊断二人的证候，然后分别开了两个药方，嘱咐说：倪寻应当下泄，李延则须发汗。两人心中纳闷，不解地问：我们的病症一样，为何用不同的治疗方法？华佗笑着解释说：倪寻属外实证，李延属内实证，所以治疗方法应当有所区别。两人按照所开药方服下了汤药，第二天清早便觉得雾散云开，身轻如燕了。

表面上看起来，华佗治病很简单，三言两语便可妙手回春。实际上他头脑中辨证施治的思维过程是相当复杂的。诚如张景岳所言,他必须首先诊知"寇"之所在,然后选择"精兵"攻之。然而,历代医家在长期实践中摸索出来的治疗"精兵"又是多种多样的,究竟用哪一种才算是对症施治,确实不是一日之功,非有高深的医术不可。

当然,中医的治疗方法还是有章可循、有法可依的。张景岳的贡献之一,就在于从千百种治疗方法中为医家归纳整理了便于依循的基本章法,即所谓"八略"。略,是兵家术语,意思是谋略、方略。以此归纳治病方法,决非无所用心,其蕴意大概在于提醒医家吸取兵家思维方法,以作战的姿态去治病。这八略,中医学上称作八法,即汗、吐、下、温、和、补、清、消八种治疗方法。

**汗**,即发汗,是基于"其在表者,汗而发之"(《素问·阴阳应象大论》)理论,开泄腠里、解除表邪的治法。外邪入侵,最初起于皮毛,以后由表入里,逐渐深入。邪在皮毛的时候,用汗法来宣发肺气,调畅营卫,开泄腠里,促使人体溱溱汗出,进而迫使在肌肤表面的外感六淫之邪随汗而解。但一般认为,当病人大失血、剧烈呕吐、腹泻时,要慎用汗法。

**吐**,即涌吐,是基于"其在高者,引而越之"(《素问·至真大要论》)理论,涌吐痰涎、胃中食积及毒物的治法。凡是痰涎壅塞在咽喉,或顽痰蓄积在胸膈,或宿食停滞在胃脘,或误食毒物尚留在胃中未下等,都可及时用吐法使之涌吐而出。由于吐法能引邪上越,宣壅塞而导正气,所以在吐出有形实邪的同时,往往汗出,使在肌表的外感病邪随之而解。吐法多用于实邪壅塞,病情急剧的病人。至于年老体弱者,尤其是孕妇,则应忌用。

**下**,即下泻,是基于"病在里,则下之而已"(《素问·至真大要论》)理论,通便除积、荡涤实热、攻逐水饮的治法。凡邪在肠胃,而致大便不通,燥屎内结,或热结旁流,以及停痰留饮,瘀血积水等邪正俱实之证,均可使用。年高津液枯耗者及孕妇尽量少用。

**和**,即和解,是基于"伤寒邪在表者,必渍形以为汗;邪气在里者,必荡涤以为利。其于不内不外,半表半里,既非发汗之所宜,又非吐下之所对,是当和解则可以矣"(《伤寒明理论》)理论,通过和解或调和的作用以扶正祛邪为目的的治法。

**温**,即温煦,是基于"寒者热之","治寒以热"(《素问·至真大要论》)理论,回阳救逆、温中祛寒、温通血脉的治法。凡脾、肾阳虚,或者受寒邪侵袭,都可用温法治疗。

**清**,即清热,是基于"热者寒之","温者清之"(《素问·至真大要论》)理论,清热解毒、清气凉血的治法。凡是热证,都可用清法。

**消**,即消散,是基于"坚者削之","结者散之"(《素问·至真大要论》)理论,消食导滞、消瘀散结、化痰化湿的治法。消法所治,主要是病在脏腑、经络、肌肉之间,邪坚病固而来势较缓,且多虚实夹杂,尤其是气血积聚而成之症,不能迅速消除,必须渐消缓散。

**补**,即滋补,是基于"虚则补之"(《素问·三部九候论》)理论,补益精血、培补元气的治法。凡人体气、血、阴、阳不足,导致脏腑出现各种虚证,都可用补法治之。

清代医学家程锺龄在《医学心悟》中指出:"一法之中,八法备焉。八法之中,百法备焉。"在他看来,治疗方法自有体系,上述八法是这个体系中的八条主线,而每条主线上又都连接着若干条支线。作为一个医生,应当具备这种体系的概念,并能够随时从中选择对症施治的方法。至于如何选择治疗方法,他进一步强调说:

"病变虽多,而法归于一。"这里的"一",大概就是"一心",与兵家所说的"运用之妙,存乎一心"意思相同。通俗地说,病情虽然千变万化,但是只要平时在心中对各种治疗方法都能融会贯通,临床施治时便能运用得巧妙得当。

## 22、排兵布阵，用药组方

明代医学家张景岳把军事上排兵布阵的指导思想用到医学上的用药立方，不愧是借它山之石磨己之刃的绝妙联想，贴切新颖，形象生动，令人击掌叫绝。但是，要通晓个中的深意，不能不费一点笔墨。

阵法，就是排兵布阵的方法。先秦兵书《司马法》说得对："凡战，非阵之难，使人可阵难。非使可阵难，使人可用难。"这就告诉我们，布阵不过是排列组合出来的各种战斗队形，关键是要使每个士兵能够在阵形中发挥作用。特别是开战之后，又要根据敌情、我情、地形、态势的变化，调整阵形，克敌制胜。

我国古代兵书《六韬·均兵》说："战则一骑不能当步卒一人。三军之众成阵而相当。则易战之法：一车当步卒八十人，八十人当一车；一骑当步卒八人，八人当一骑；一车当十骑，十骑当一车。"这就是说，在平原地域上作战，如果一名骑兵同一名步兵单个战斗，骑兵打不赢。如果组成方阵，则骑兵可同八倍于己的步兵作战。

张景岳无疑是从阵法中悟到了排列组合的奥妙，他将前人的药方1516首分列为"古方八阵"，将当时的新方186首分列为"新方八阵"。按处方的不同思路和用途，八阵的名称分别是补、和、攻、散、寒、热、固、因。

**补阵**，以补为主，适用于元气亏损，体质虚弱者。凡气虚者，宜补其上，可用人参、黄芪等药；精虚者，宜补其下，可用熟地、枸杞等药；阳虚者，宜补而兼温，可用桂、附、干姜等药；阴虚者，宜补而兼清，可用门冬、芍药、生地等药。总之，气因精而虚者，自当补精以化气；精因气而虚者，自当补气以生精；善补阳者，必于阴中求阳，则阳得阴助而生化无穷；善补阴者，必于阳中求阴，则阴得阳升而泉源不竭。

**和阵**，以调和脏腑阴阳的偏胜为目的，或者和解表里、上下、寒热。凡病兼虚者，补而和之；兼滞者，行而和之；兼寒者，温而和之；兼热者，凉而和之。凡阴虚于下，而精血亏损者，忌用利小水之药，如四苓通草汤等。阴虚于上，而肺热干咳者，忌用辛燥之药，如半夏、苍术、细辛、白术等。阳虚于上者，忌用消耗之药，如陈皮、砂仁、木香、槟榔等。阳虚于下者，忌用沉寒之药，如黄柏、知母、栀子、木通等。总之，以调平元气，不失中和为主要目的。

**攻阵**，以攻为主，适用于急性病、实证。这些病症往往邪固疾深，势如强寇，应当迅速讨伐。凡攻气者，攻其聚集之处，可使气机通畅。攻血者，攻其瘀积之处，可使血液流通。病在阴者，勿攻其阳；病在里者，勿攻其表。虽然用攻之法可以祛疾除病，但也往往给未受邪之处带来损伤，实属不得已而为之。所以，处方用药必须慎用此法。

**散阵**，以散发为主，主要用于驱散表而病邪。"邪在肌表，当逐于外，拒之不早，病必日深"。用散之方，应当熟知药性。如黄麻、桂枝消散有力，防风、荆芥、戴苏消散平缓，细辛、白芷、生姜消散温和，羌活、苍术能走经去湿而消散，柴胡、干葛、薄荷具有凉散作用，等等。

**寒阵**，主要治疗热证，用泻火方药，或者用生精养阴之药。大凡清凉之物皆能泻火，但必须分清其药性轻清重浊。轻清者可以清上，如黄芩、石斛、天花等。重浊者可以清下，如栀子、黄柏、龙胆等。

**热阵**，与寒阵相对，主要针对寒证，用温热药来助阳祛寒。如干姜能温中也能散表，呕吐无汗者可用之。肉桂能行血，善达四肢，血滞多痛者可用之。吴茱萸善暖下焦，腹痛泄泻者用之极妙。肉豆蔻可温脾胃，餮泄滑利者用之最有奇效。

**固阵**，用固涩的方药来治疗滑泄不禁之类的疾病。如久嗽为喘而气泄于上者，宜固其肺。久遗成淋而精脱于下者，宜固其肾。小便不禁者宜固其膀胱。大便不禁者，宜固其肠藏。汗泄不禁者，宜固其皮毛。血泄不止者，宜固其营卫。凡因寒而泄者，当固之以热。因热而泄者，当固之以寒。总之，在上者，在表者，皆宜固气，气主在肺。在下者，在里者，皆宜固精，精主在肾。

**因阵**，是因证立方的意思。如痈毒之起，可设法消肿。蛇毒之患，可设法解毒。汤火伤及肌肤，可设法散热。跌打伤其筋骨，可设法续接。这些情况下，都必须因证而用药。

张景岳排列药方八阵，十分注重对症下药，分析了许多药物的性质和用途，以上仅仅是取其大意而已。与治疗八略相应，张景岳反复强调，药方八阵也应当互相配合，灵活变化。清代医学家徐大椿对此作了形象的说明，他在《医学源流论》中说："若夫虚邪之体，攻不可过，本和平之药，而以峻药补之；衰敝之日，不可穷民力也。实邪之伤，攻不可缓，用峻厉之药，而以常药和之；富强之国，可以振威武也。"意思是说：如果体弱病久，用药不宜过猛，可以和平之药为主，而以烈性药为辅，犹如国家衰弱之时不可过分消耗民力一般。如果体壮初病，用药不可迟缓，当用烈性药强攻，而以平和之药调和，犹如国力富足强大，可以出兵作战一般。这其中既强调虚实相变，又强调攻和相依，道出了药方八阵的奥妙之所在。

## 23、正合奇胜巧应变,君臣佐使须分明

明代医学家徐春甫曾在《古今医统大全》中比喻说:"治病犹对垒。攻守奇正,量敌而应者,将之良;针灸用药,因病而施治者,医之良也。"这个比喻颇为恰当,抓住了治病与对垒都必须根据情况而灵活变换对策的共同特点。他所说的"攻守奇正",相当于我们今天所说的作战方法。《孙子兵法》有句名言:"凡战者,以正合,以奇胜。"意思是,作战对垒,通常的打法应当是用主要兵力打敌正面,用次要兵力打敌翼侧。对这一常法,它还具体分析说:"十则围之,五则攻之,倍则分之,敌则能战之,不若则能避之。"我们以"五则攻之"为例,加以说明。孙子的意思是,如果我五倍于敌,可以采取主动进攻的打法。在战术上,那就应当以三倍于敌的主要兵力打敌正面,以两倍于敌的次要兵力打敌翼侧,而不可相反。

徐春甫把这一兵法原则应用到医疗上来,认为"针灸用药,因病而施治",正同"量敌而应"、因情用兵是一个道理。

我们知道,中医用药组方讲究"配伍"。所谓配伍,就是按照病情的需要和药物的性、味、归经,有选择地将两种以上药物配合在一起应用。这一理论最早形成于《神农本草经》之中。作者把书中所收录的365种药物分为上中下三类,同时又将这三类分别取名为君、臣、佐、使,以比喻各种药物在处方中的主次作用。"君"即指处方中起主要作用的药;"臣"是指帮助主药以加强其功效的药;"佐"是指起辅助作用、减轻主药毒性或作反佐药用的药;"使"是指引药直达病所的药。书中归纳了每一类药的主要作用,并详细说明了每一种药的性、味、归经及生长之地,从而为医家分析药物特

性,合理组织配伍,提供了明确的依据。

显而易见,"君"药相似于"正兵",是打主攻的部队;"臣"和"佐"、"使"相似于"奇兵",是打助攻的部队。明白了这一点,我们不能不佩服徐春甫的类比是何等的恰切。其实,最令人叫绝的是,在治病用药上这种"正合奇胜"的战术思想确有生动具体的表现。

中医的"正治""反治"思想正可与兵法的奇正多变、因敌制胜思想双双比美。

所谓"正治",即指直接针对疾病的性质、病机从正面治疗。因药性与病性相逆,为治病之常法,叫做"正治"。例如:

**热药能祛寒**,用以治疗寒性病证,即所谓"寒者热之";寒性病大体可分两类,一是外寒,如外受风寒而成的风寒感冒之类。二是人体某个脏腑甚至全身机能衰退,出现身冷、腹泻、浮肿、阳痿等症状,称为里寒。能治疗寒性病的药,多有辛味,其性质必是温热。如苏叶、葱白辛温,能治风寒感冒。干姜辛热能治胃寒疼痛,成药理中丸就以干姜为主要成份。

**寒药能清热**,用以治疗热性病证,即所谓"热者寒之";热性病也可分为两大类,一是表热,如感冒病人口干、咽痛,称为风热感冒,其病在表。二是里热,表现为身热、口苦、心烦、牙龈肿痛、便秘、尿黄等症状。能治热性病的药,多有苦味,性质必定寒凉,如治风热感冒的桑菊感冒片,其中桑叶、菊花都有苦味;能治心烦、牙龈肿痛、大便秘、小便黄的黄连上清丸,其中黄连味苦而性寒。

**补药能滋养**,用以治疗虚弱病证,即所谓"虚则补之";虚弱之病种类很多,例如心慌胆怯、失眠健忘,属于心虚,可用天王补心丹治疗。面色萎黄或苍白,气短乏力,大便稀溏属脾肺虚,可以用参苓白术丸治疗。腰疼腿软,四肢发凉,阳痿等症,属于肾虚,可用金匮肾气丸治疗。上述能治各种虚弱之病的药物,都属补药之类。

**攻药能泻下**,用以治疗属实的病证,即所谓"实则泻之";实性病是指人体内存在着"多余"的成分,如一切致病素因,无论细菌、毒物等侵入人体,都是多余的;饮食物不能消化和正常排泄,停留在体内,也是多余的,如燥屎、尿潴留;人体原本正常的气血等,由于某些原因引起紊乱,不能正常运行,也变成有害的多余之物了,如瘀血等。这些多余物统称"实"或"实邪",都应该被排泄出去——如发汗、利尿、通便、破瘀血等,从治疗法则来说,就叫"实则泻之"。

又如,用具有下降作用的药物治疗气机上逆的病证,用化痰的药物治疗痰证,用消食导滞的药物治疗积食和瘀血证,用驱虫、杀虫药物治疗虫证等等,都属于正治法。由于临床上多数疾病所表现的征象与疾病的性质相符,因而正治法乃是临床上最常见的一种治疗方法。

战术因敌情而变化,用药因病情而变化。如果在疾病出现假象时也可以采用相反的方法治疗,这就要谈到反治。

例如,某些外感热病,在其里热盛极之时,有时可见到四肢厥冷的寒象。因其寒象是假,而热盛才是本质,所以仍然必须用寒凉药物进行治疗,此即所谓"寒因寒用"。

某些寒性病证,时有反而出现面颊浮红,身热,口渴,烦热不安等热象。因其热象是假,而寒盛才是其本质,所以仍然必须用温热药进行治疗,此即所谓"热因热用"。

又如,某些脾虚之病,反而出现大便阻塞不通的假象,治疗时仍用补脾药,补药有补充、填塞之意,病症已见

"阻塞'",治疗也有"填塞",这就叫"塞因塞用"。

此外,由于食积停滞,影响运化所导致的腹泻,则不仅不能用止泻药,反而应当用通下消滞的药,使肠胃通和,则泄泻自止。因消滞药有通利的作用,而病症有通利的现象,此即所谓"通因通用"。

以上"寒因寒用"、"热因热用"、"塞因塞用"、"通因通用"几种治疗方法,不同于"以正合"的正治之法,表面上"顺从"疾病症状的假象,所以中医理论上将这类治疗方法称作"反治"、"反佐"。但是,就其本质来说,仍是针对疾病内在本质而施治的方法。

"用药如用兵",古代医学家的这一观点不仅是一种贴切的比喻,而且具有丰富的内涵和深刻的哲理。

## 24、用兵宜慎，服药戒躁

有这样一个故事：

一位姓张的人患有痞证，胸中积结不通，饮食皆阻塞难下，不得已而求治于医生。医生诊断之后说，此病非用下法疏通内脏不可，并给他开了一些药。张氏回家之后，心想药可祛疾，何不多吃一点，以便一举赶走病魔。于是，他将医生所给的药一饮而下。果然，不出一日，胸中堵塞的感觉很快就消失了，并且饮食顺畅，呼吸自然，恍若未曾病过一般。不料，几天之后旧病复发。张氏又象上次那样，一口气服下比前次更多的药，效果果然更加明显。此后的一个多月里，痞证连续发作了五次，每次都很快被猛药所压服。然而，张氏也因此而大伤元气，呼吸困难，虚汗不止，浑身颤抖，没有做什么消耗体力的劳动，却总是感到疲困难支。张氏心中纳闷：为什么这个毛病老是来而又去，去而又来呢？

后来，他听说楚国南部有一位良医，便前往就诊。医生叹息地说："你不要抱怨疾病顽固不化，其实这都是你不正确的服药方法所造成的。你的病是由于脏腑功能失调，气与血不通畅而引起的痞证。病症横阻于胸中，危害较大。对此病症如果集中用大量的药物猛攻，自然不须片刻便能有压制的效果，但是体内的和气承受不了这样的打击。所以，每次从表面上看起来你的病症是治好了，其实你体内正常的和平之气也随之而受到了损伤。如此多次反复，当然会不劳动而虚汗不止，不行走而浑身战栗，终日疲困乏力。为了能够既治疗痞病又不至于损伤体内的和气，你先回去好好休息三个月,然后再来取药医治。"

三个月后,张氏如期来找楚国的医生。医生说:"你的元气有所恢复,可以开始服药了。"同时,他还嘱咐道:"按规定的量服用,三个月之后疾病开始缓解,再过三个月之后身体开始康复,一年之后便可恢复正常。千万不可亟进。"张氏按其嘱咐所行,果然逐步见效。

病愈之后,张氏拜谢楚医,并请教其治法的奥妙。医生没有从医学上正面回答,而是分析了秦国实行严刑苛政以致早亡的教训,由此得出一切事物的共同规律:凡急于求成的,最终总会有所损伤,而要想最终得到完美的结果,在起始之时就不要有一蹴而就的思想。

这则故事见于宋朝张耒的《药戒》,它道出了一个很深刻的含义,那就是"慎疾慎医"。

《论语》记载:"子之所慎:斋、战、疾。"孔老夫子在三件事情上坚持慎重态度,一是祭祀神灵祖宗,二是战争,三是疾病。这是很有道理的。中国古代的兵家一贯认为,兵凶战危,"一人之兵,……天下皆惊"(《司马法·仁本》),不可轻举妄动。因此,兵学家们无不主张慎兵慎战。

不言而喻,慎疾与慎战在思维方法上完全是两个相互吻合的符契。所以,《本草类方》指出:"用药如用刑,误即便隔死生。"

患病的人及其亲属对于疾病大多有一种畏惧恐慌的心理,唯恐危及性命。因此,总是希望寻求良医,吃到好药,尽快痊愈。可是,大凡药物,总有一定的副作用,必须慎重对待。

药物剂量的大小并非随心所欲的,主要应从三个方面把握:一是根据药物性能确定用量。凡有毒、峻烈的药物,用量宜小,并应从小量开始,根据病情需要再考虑逐渐增加,不要过量。一旦病势减退,应逐渐减量或立即停服,以防中毒或产生副作用。一般药物,质重的如矿物、贝壳类药,用量宜大;质轻的如花、叶类,以及芳香走散的药物,用量宜轻;厚味滋腻的药物,用量可稍重。二是根据配伍、剂型确定用量。一般说来,同样的药物,一味单用剂量较入复方为重;入汤剂用量较做丸、散剂为重;在复方中的主药用量较辅助药为重。三是根据病情、体质、年龄确定用量。一般重病、急性病及病情顽固的,用量宜重;轻病、慢性病用量宜轻;病人平素体质壮实的用量宜重;年老体弱、妇女儿童用量宜轻;新病用量宜重,久病用量宜轻。各年龄期用药剂量可参照下列表格:

| 年 龄 | 剂 量 |
| --- | --- |
| 初生—1个月 | 成人量 1/18 — 1/14 |
| 1 — 6 个月 | 成人量 1/14 — 1/7 |
| 6 个月 — 1 岁 | 成人量 1/7 — 1/5 |
| 1 — 2 岁 | 成人量 1/5 — 1/4 |
| 2 — 4 岁 | 成人量 1/4 — 1/3 |
| 4 — 6 岁 | 成人量 1/3 — 2/5 |
| 6 — 9 岁 | 成人量 2/5 — 1/2 |
| 9 — 14 岁 | 成人量 1/2 — 2/3 |
| 14 — 18 岁 | 成人量 2/3 — 全量 |
| 18 — 60 岁 | 全量—成人量 3/4 |
| 60 岁以上 | 成人量 3/4 |

服药的方法也要讲究,也要慎重,否则便会带来不良后果。

一般补养药宜在食前服;对胃肠有刺激的药物宜在食后服;杀虫药及泻下药宜在空腹时服;治疟药宜在发作前服;安神药宜在睡前服。急性病应立即服药,不拘时间;慢性病服丸、散、膏、酒者应有定时,如1日3次或每日早晚各1次,以使药物持续发挥治疗作用。

无论食前或饭后服药,都应在饭前后 1~2 小时左右,以免影响疗效。一剂中药,通常分 2~3 次服用。病缓者可早晚各服 1 次,病重病急者可每隔 4 小时服药 1 次,使药力持续,利于顿挫病势。在应用发汗、泻下等药时,若药力较强,须注意患者体质,一般以得汗、下为度,中病即止,不必尽剂,以免汗、下太过,损伤正气。所以,古代医家强调:"夫药无次序,如兵无纪律,虽有勇将,适以勇而偾事"(李杲《珍珠囊指掌》)。为了避免重蹈前文所说张某人的覆辙,古代医学家的慎疾慎药思想是值得重视的。

合理用兵,则利国利民,而穷兵黩武,却是自取败亡之道。用药适当,则能祛病强身,而孟浪用药,则使气血衰败。因为凡药都有阴阳之偏,正确使用可以纠正人体的阴阳失调而治病,但如矫枉过正,必然伤害正气。怎样衡量用药是否适当呢?《黄帝内经》做了如下的规定:

"大毒治病,十去其六;常毒治病,十去其七;小毒治病,十去其八;无毒治病,十去其九;谷肉果菜,食养尽之。"

这里的"毒"是指药物作用剧烈而言。尽管是使用和平无毒之药,也只应去掉疾病的九成,便应停药。余下的微小之病,可以通过饮食调养和身体锻炼,使正气自行恢复而痊愈。即使"补品",虽然可以增强人体正气,但服用过久、过量同样会导致体内阴阳失调而生病。诚如《素问·至真要大论》所说:"久而增气,物化之常也;气增而久,夭之由也。"

## 25、攻敌先攻心,治病先治心

在中国古代医学史上,"心"一直是一个举足轻重的概念,并且很早就有血肉之心与神明之心的区分。血肉之心主五脏六腑及全身血脉;神明之心主精神、意识及思维。如《黄帝内经》中说:"心者,五脏六腑之大主也,精神之所舍也。"明代的《医学入门》也指出:"有血肉之心,形如未开莲花,居肺下肝上是也。有神明之心……主宰万事万物,虚灵不昧者是也。"相形之下,先人们对神明之心的认识可能更早、更为重视。

产生于公元前八世纪左右的语录式兵书《军志》中曾有"先人有夺人之心"的名言,认为先发制人者应当首先打击敌人的军心士气。由此可知,当时的人们不仅已经认识到心具有主宰人的精神和意识的作用,而且认识到人的精神和意识对战争胜负有着直接的影响。春秋时期,兵圣孙武继承了这一思想,明确提出"三军可夺气,将军可夺心"的主张,并论述了一系列基本方法,诸如:"上兵伐谋,其次伐交,其次伐兵,其下攻城",要求首先斗志斗谋,从心理上打乱敌人的决心和部署,然后再一举消灭之;"能而示之不能,用而示之不用,远而示之近,近而示之远",强调通过假象造成敌人心理上的错觉;"先夺其所爱,则听矣",主张抓住敌人心理上最关注之点,以便牵着敌人的鼻子走。在漫长的历史长河中,随着《孙子兵法》的流传,这些方法成为后世兵家克敌制胜的法宝。三国时期马谡的攻心之策就是具体运用这些法宝的产物。

公元225年,诸葛亮进军南中。后主差马谡前往犒军,分发后诸葛亮问道:"吾奉天子诏,削平蛮方,久闻幼常(马谡名)高见,望乞赐教?"马谡认为,南中是少数民族地区,人民骠悍骁勇,加以地险道远,很难彻底战而胜之。即使一时使之屈服,不久之后也可能重新成为祸患。而且,对于蜀汉政权来说,主要对手是北面的曹魏和东面的孙吴,为避免腹背受敌,对于作为后院的南中尤须注意和抚。因此,他建议说:"夫用兵之道,攻心为上,攻城为下,心战为上,兵战为下,愿丞相但服其心足而已。"诸葛亮颇有同感,完全采纳了这一建议。在南征过程中,诸葛亮刚柔相济,恩威兼施,最终降服了南中首领孟获。

清代医学家徐大椿曾在《医学源流论》中总结说:"孙武子十三篇,治病之法尽之矣。"的确如其所言,孙子的攻心之法在医学领域也有着广阔的用武之地。所不同的是,兵家攻心,旨在扰乱敌人的正常心理状态;医家攻心,重在恢复患者的正常心理状态。

《黄帝内经》指出:"精神不进,志意不治,故病不可愈。"意思是说,一个医生治病,如果只考虑生理与病理的变化,而不考虑精神的,即心理的变异,从心理上、精神上配合治疗,疾病是不可能治好的。明代医学家李中梓在《医宗必读》中强调:"境缘不偶,营求未遂,深情牵挂,良药难医。"在他看来,由于心情不佳而造成的种种病变,单靠药物治疗是无济于事的。既然"心病还须心药治",所以历代名医都主张"善医者,必先医其心,而后医其身",高度重视心理因素在治疗中的能动作用。

心理治疗是指不用针灸、药物、手术等有形的治疗方法,而借助语言、行为等治疗手段对患者进行启发、开导、安慰及调理,甚至刺激,以提高病人对疾病的认识,解除其顾虑,增强其战胜疾病的信心和能力。《三国志·魏志·华佗传》中就有华佗利用心理刺激治疗疾病的记载。

书中记述:有一位郡守长期患病不愈,华佗诊断后认为,此人若能大怒一场则可能疏通胸中积疾,驱除病魔。于是,他故意索取患者的钱财而不给予任何治疗,不久之后又不辞而别,甚至留下一封书信辱骂一番。郡守忍无可忍,大发雷霆,派人追捕华佗,试图杀之以解心头之恨。郡守的儿子知道华佗的用意,悄悄拦住奉命追捕的差人,并让其谎称华佗已逃之夭夭。郡守得知此讯之后,怒火中烧,血涌心头,不禁连吐黑血数升。出乎其意料,黑血吐出不久,宿疾也就不知不觉地痊愈了。

现代医学研究发现,人体免疫系统受神经——内分泌系统调节,因而也受到认识、情绪等心理因素的影响。不良情绪状态可导致明显抑制人体的免疫功能,使免疫细胞的自我识别能力下降、吞噬细胞的功能减退,使机体内平衡受到破坏、协调有序的生理机理产生紊乱。癌细胞的发生和发展就是这种影响的主要结果之一。所以,对于癌症病人尤须进行心理治疗。

我国当代著名电影表演艺术家秦怡,曾先后生过四次大病,做过七次手术,灾难屡次在她的头上降临。1962年,她因患甲状腺瘤,做了一次手术。四年之后她又得了肠癌。这是一种令人生畏的恶性病。面对死神的威胁,她没有消沉,而是理智地调节自己的心身,恪守始终保持乐观、充满活力的信条,并以顽强的毅力坚持进行养生锻炼。三十多年过去了,她不但没有被病魔所吓倒,反而成功地度过了一次又一次大手术,战胜了病魔,被人们誉为"抗癌明星"。

古今医疗实践一再证明,无论是心理治疗,还是心理调养,都有助于"不战而屈人之兵",不药而治。

与兵家心理战法一样,医家临床采用的心理治疗方法也是多种多样的。结合古今名家的经验,主要有以情胜情、以疑释疑、劝说开导、移情易性、暗示解惑、顺情从欲、气功引导等方法。

## 26、抑情制怒，以情制情

中国古代兵法中，对于将帅品德一个最重要的要求就是必须具备高度的克制能力，特别是在重大战略决策上必须抑情制怒，不为敌人的激将法而一触即跳，鲁莽行动。《孙子兵法》明确指出："主不可以怒而兴师，将不可以愠而致战。合于利而动，不合于利而止。怒可以复喜，愠可以复悦，亡国不可以复存，死者不可以复生。故明君慎之，良将警之。"这段话的意思是说，国君不可因一时愤怒而发动战争，将帅不可因一时气忿而出阵求战。符合国家利益才用兵，不符合国家利益就停止。愤怒还可以重新变为欢喜，气忿还可以重新变为高兴；国亡了就不能复存，人死了就不能再生。所以，对于战争，明智的国君要慎重，贤良的将帅要警惕。

显而易见，面对关系国家民众生死存亡的军国大事，绝不能以个人的情感去指导战争。因此，国君和将帅在任何时候都必须保持理智和冷静。

不良的精神状态和心理情绪，不仅不利于指导战争，也不利于人们自身的健康。因此，学习兵家优良的品格修养对于自身健康也是善莫大焉的。

《黄帝内经》指出："百病生于气。怒则气上，喜则气缓，悲则气消，恐则气下，惊则气乱，思则气结。"可见，情绪波动剧烈或持续过久都可能影响机体脏腑气血的生理功能，导致疾病发生。中医理论认为，喜、怒、忧、思、悲、惊、恐七种心情，既可因偏胜而致病，又可因彼此调济而治病，所以有"以情胜情，以情制情"治疗方法。这种治疗方法主要通过刺激和调节心理状态，从而抑制或排遣不正常情绪,最终控制和战胜疾病。古代医书中运用这一治法而取得成功的病例比比皆是,不胜枚举,从下述两个病例便可略见一斑。

据《名医类案》记载,某县一个差役押送犯人,用铁索将犯人缚住,行至中途犯人突然投河而死。犯人的亲属闻讯后,急忙到县衙门告状,说犯人是因差役诈骗钱财威逼而死的。衙门裁决,差役虽无任何罪过,但须赔偿一些钱财。从此,差役忧愤成疾,整天如醉如痴,胡言乱语,形同白痴一般。家人请来一位名叫汪石山的医生为其诊治。汪医生认为患者系破财而忧,必得大喜才能祛其疾,药物治疗是无济于事的。于是,他采用"喜胜忧"的方法,让患者家人用锡伪做银锭数枚,趁患者熟睡时放在其身边。当患者醒来时,猛然发现如此多的银锭,欣喜非常,爱不释手。此后,病情不知不觉地减轻,不久便痊愈了。

据《儒门事亲》记载,有一位妇女,一次外出旅行住宿在客店的楼上,偏巧当天晚上一群强盗闯进客店抓人放火,她因惊吓而从床上摔到地下。从此以后,每听到什么响动,她就惊厥昏倒而不省人事。她家里的人不得不轻手轻脚,以免弄出响声而惊吓了她。这样,一年多过去了,病也没好。许多医生都诊断为心病进行治疗,服以人参、珍珠和定志丸,但都没有什么效果。一位名为张戴人的医生诊察后,认为惊则伤胆,治疗此病重在为其壮胆。因此,他让二名侍女抓住患者的两手,分别按在高椅上,并在患者面前放一个小茶几,开始实施壮胆之术。张医生先请患者注视茶几,趁其不备时,突然用木棒猛击茶几,患者大吃一惊,险些昏倒。待患者恢复平静后,张医生问道:"我敲打茶几,你怕什么?"不久之后,他又故伎重演,并反复多次。每反复一次,患者的惊吓程度都减轻一点。然后,他又用木棍敲门,并让人悄悄撩开窗帘做鬼脸,直至患者习以为常为止。此后,这位妇女就是听到巨雷之声也不再惊厥了。

有人问道:"这是什么治法?"张戴人笑着回答:"《黄帝内经》说,受惊的,可使之平息下来。平,就是平常的意思。人们对司空见惯的事物往往不会产生惊怕的心理反应。如今我用的实际上是思胜恐的方法,通过反复重现曾经使患者害怕的声响,使她认识到原来的害怕是多余的,从而克服了恐惧心理。

除"喜胜忧"、"思胜恐"外,"悲胜怒"、"恐胜喜"、"怒胜思"等,都属于以情胜情的范围,其基本原理都是用正常的心理压制或代替非正常的心理,从而祛除疾病。

## 27、乖其所之,移情易性

　　古今中外的战争历史证明,要想在战场上立于不败之地,掌握调动敌人而不被敌人所调动的主动权是十分关键的因素之一。独具慧眼的孙子自然不会忽略这一点,他在《孙子兵法》中以浓墨重彩的笔调阐述了"致人而不致于人"的观点,并提出了一系列巧妙的方法。其中一个颇为奇巧的方法是"乖其所之"。乖,即违背、相反,引伸为改变、调动的意思。孙子认为,如果我不打算与敌人交战,即使画地设防,敌人也无法来同我作战,这是因为我设法改变了敌人进攻方向的缘故。

　　孙子的后代孙膑可谓深得其旨,并运用这一方法导演出我国战争历史上十分精彩的一幕——围魏救赵。公元前354年,魏国围攻赵国都城邯郸(今属河北),次年赵王遣使求救于齐国。齐王命田忌、孙膑率军前往救援。孙膑考虑到魏的精锐部队在赵国城下,内部空虚,便引兵攻打魏国都城大梁(今河南开封),诱使魏将庞涓兼程赶回应战。庞涓不知是计,仓促回师救援。孙膑又在中途设伏袭击,一举大败魏军,生擒魏将庞涓。

　　明代医学家张景岳曾经说过:"任医如任将,皆安危之所关。……第欲以慎重与否观其仁,而怯懦者实似之;颖悟与否观其智,而狭诈者实似之;果敢与否观其勇,而猛浪者实似之;浅深与否观其博,而强辩者实似之。"(《景岳全书·传忠录下》)这段话虽然是就用将而言的,其实"斗将"也不外乎于此。在两军相交的战场上,首先必须从多方面透过表面现象准确把握对方将领的心理素质,然后再针对其特点采取相应的策略。孙膑就是因为摸透了庞涓狂妄自大、骄傲轻敌的心理特点,因而能够略施小计便打乱其战略决心,乖其所之,使其陷入顾此失彼、进退失据的境地。

那么,治病祛疾的过程中能不能采用"乖其所之"的办法呢?答案无疑是肯定的。中医有一种移情易性的心理治疗方法与"乖其所之"的心理战法可谓原理一致。

所谓"移情",是指分散病人对疾病的注意力,或改变病人内心虑恋的指向性,使其从某种情感纠葛中解放出来,转移至别的人或事物上。所谓"易性",是指通过某种手段排除病人内心杂念,或改变其错误认识和不健康习性。历代中医的实践证明,这一方法不仅是简便的,而且是有效的。《三教九流的传说》一书中收录的一则故事就生动地证实了这一点。

古时候有一张一王两医生,张医生自认医术高明而看不起王医生,然而求王医生看病的人总是络绎不绝,张医生却门可罗雀。张医生眼看王医生的药铺日益红火,心中十分嫉妒,气得整日茶饭不思,坐卧不安,以致病倒在床。请来看病的医生难免都要问起他药铺的生意如何,真是"哪壶不开偏提哪壶",使他气上加气,不但服用的名贵药不见效,病情反而与日俱增。万般无奈,张医生的儿子只好对父亲说:"还是请王医生来看看吧,听说他的医术很神奇。"张医生一听,更是气不打一处来,大声吼道:"我的病就是他给气的,让他给我看病,你是嫌我死得慢啊!"

然而,随着病情不断加重,张医生最后还是迫不得已同意请王医生来诊治。王医生向来受张医生的嫉妒和排挤,知道其病因所在。于是他煞有介事地用手按着床帮(而不按其手)给张医生"号脉",然后郑重地说:"你的病是经血不调,我开贴药给你吃,保证药到病除。"张医生接过药方一看直摇头。王医生马上说:"你是怕吃药苦吧,不吃药也行,贴上一张膏药吧!"说着便掏出膏药点灯就烤。张医生露出脊背准备贴药。王医生连声说："盖好，盖好"，同时一巴掌将膏药贴在了张医生床头的墙上，然后拱手说："张大哥的病今晚就见轻，明天能下床，后天可出户，七天便好透，小弟告辞啦。"

王医生走后，张医生哈哈大笑地说："他王医生原来如此狗屁不通。号脉哪有按着床都的？经血不调是女流之病，我一个老头岂能得这等毛病？就算经血不调，应该开当归、白药、红花、丹皮之类的药，他开的却是下奶的药。再说，给病人贴膏药哪有贴到墙上的？这真是天大的笑话！"此后，他逢人便说，边说边笑，甚至手舞足蹈地比比划划，显得特别开心。果然不出王医生所言，七天之后，张医生胸中的恶气悄然消失，坐卧自如，饮食大增。后经旁人点破，他才恍然大悟，原来王医生采用的是移情易性之法，从此对王医生心服口服。

1984年《长寿》杂志上登载的一则移情易性而治愈疾病的故事也是很有说服力的。故事说：当代著名的儿童教育家孙敬修先生，童年时代因与同伴较量手臂力量用力过猛造成内伤，以致长期吐血，并且每到凌晨三点就吐血。他母亲既无钱为他求医买药，也无力买什么营养品给他调养。眼看着他的体质一天不如一天，母亲忧心如焚。有一天，母亲守护着他，眼瞅着挂钟又要敲响三点了，情急之中她把钟拨快了一小时。一会儿之后，孙敬修醒来了，瞧瞧钟已经四点了，就问妈妈为什么还没有吐血。妈妈灵机一动，回答说："孩子，你的病好了。你看，现在都四点了，还没有吐血。"从此，孙敬修的心理障碍逐步消除，病也渐渐地好了起来。直到八十多岁，他还经常在电台和电视节目中给小朋友讲故事。

这些故事说明，对于"内伤七情"为主的疾病，只要象军事上用将、斗将那样,准确地把握引起疾病的原因及患者的心理状态,设法排遣不正常的情思,改易引起疾病的心志,就有可能治愈疾病,产生与"乖其所之"异曲同工的效果。而这种"移情易性"法治病的道理,在《黄帝内经》中早有解释。该书中有一篇文章的题目就叫做"移精变气"。是说人之所以有病,是由于各种原因引起了体内的"气"发生紊乱,有时可以通过转移病人精神(移精)的方法,来改变其紊乱之气,使之恢复正常(变气),从而治愈疾病。《黄帝内经》的另一篇文章《杂病》,还举出"哕"病,即膈肌痉挛,因为病情较轻,可以用"大惊之"治愈它。这是由于"惊则气乱",以乱治乱,待其心情平静下来之后,"气"从新调整过来,"哕"自然会好了。

## 28、兵不厌诈，疑病疑治

疑心病，是人们日常生活中较为多见的毛病。在中医学理论中，许多医学家也将多疑视作多种疾病的诱因之一。他们认为，有的人患某种疾病以后，容易产生各种各样的怀疑或猜测，或小病疑大，或轻病疑重，或久病疑死。甚至有的人本来没有什么病变，由于偶然受到某种内外刺激，就疑神疑鬼，怀疑自己得了这样那样的重病，以致无病之躯真的疑出一场大病来。徐灵胎的《洄溪医案》中有这样一个病例：

一个名叫李鸣古的书生，家贫不得志，遂得奇疾。他每天晚上似乎听到有人骂他，但只闻其声而不见其形，咒骂的语言恶毒不堪。于是他恼恨终日，不寝不食。多方安慰也无济于事。他的叔叔何小山出于同情，陪他一起就诊。医生问诊时，李鸣古说："我没有病，只不过有人骂我罢了。"医生指出："这就是病之所在。"何小山也劝慰道："你的学问和人品是人人钦佩的，哪还会有什么人骂你呢？"岂料，李鸣古却突然哭泣起来，边哭边说："别人劝我还可以理解，叔叔你明知我有病也要来劝我，则太不近情理了。看来，昨天在隔壁骂我整整一天的人肯定是你，今天何必还要当面奚落呢？"何小山连忙解释道："我昨天从早到晚都在别处，怎么会在你隔壁？何况你隔壁是谁家我都不知道，怎么能进到他家去呢？"然而，你越辩，他越疑，无济于事。后来，李某其人终于忧郁而死。

《黄帝内经》说，忧伤肺，而肺部病变若不及时治愈又可能传到其它脏器，乃至影响五脏六腑的气机正常运行。所以，无病呻吟，自疑病笃，最终有可能导致丧生。而且，这类疾病既难以用药治疗，也难以用劝导、解释等一般的心理方法治疗。那么是不是无法可治呢？当然不是。古代兵法中的诈术，对于治疗有些疑心病堪称灵丹妙药。

"诈"，即诡诈、欺骗的意思。孙子认为作战能否取得胜利，会不会使诈术是关键因素之一，所以他的一句名言是"兵以诈立。"

事实上古代医学家也确实注意到了诈术的医疗作用。汉代医学家张仲景就曾在《伤寒论》中提出过"以诈治诈"的治疗方法。章虚谷解释其义说：如果某人"不惊而起，左右盼视，身健心清也；问其病状，三言三止，吞吐支吾，无痛苦可说也；脉之咽唾，无呻吟声，而脉自和，则可知其为诈病矣。即以危言恐之，彼畏毒药针灸，其病自愈，是以诈治诈之妙法也。"军事上的兵不厌诈，在这里移植为治病上以疑释疑。

《北梦琐言》中就记载过这样一个病案：

唐代时期，有一位妇女随丈夫去南中省亲，途中误食一虫，常常耿耿于怀，于是忧虑成疾，多次请人治疗也不见效果。后来，一位名为元桢的医生为其诊断，知其得病缘由。于是，他悄悄告诫妇人的侍女："我将要给她服用吐泻的药，在她吐泻之时，你用盘盂接住，然后谎称看见其中有一个小虾米游走，千万不能让你的女主人看出破绽。"侍女依计而行，患者果然信以为真，转忧为喜，不久之后便病容全消了。

由于心理障碍导致疑心病的案例，古代有，现代也不少见。

据说，有一位中年妇女自从参加了女友的丧礼之后，开始觉得咽喉不适，吃硬东西发哽，并且逐日加重，连喝水也咽不下去，体重从一百斤下降到几十斤,多处求医无效。后来,她慕名向一位医学教授求治。老教授经过详细检查和询问后,弄清了她发病的原因。原来,她的女友是因食道癌而死的,她由此而怀疑自己也得了这种病,于是疑惧成疾。老教授认真地对她说:"我们这里有一种特效药,是专治这种病的,已经治好了不少象你这样的病人。不过,这种药注射起来比较疼。"并嘱咐其家人为她准备一些可口的东西。这位病人接受注射时果然感到很疼,因而对教授的话深信不疑。回到家后,她立即按老教授的嘱咐吃了一碗稀饭,又吃了一些水果。令她惊奇的是,吞咽这些东西时居然没有原来那么难受了。此后,经过连续几天治疗,病就痊愈了。这位妇女对老教授感激不尽,专程登门致谢。致谢之余,她问是什么药物这样有效。老教授笑着说出三个字——"蒸馏水!"

上述病案中,医生之所以略施小计就能治愈病人,关键在于他们准确地掌握了病人心理病变的原因,用善意的欺骗巧妙地转移了病人的注意力,化解了病人的疑心,从而收到良好的治疗效果。难怪有人曾经诙谐地说:"对于病人撒谎,上帝都是允许的。"由此可见,兵法上的诈术对于治疗小病疑大、无病自扰之类的心理疾病不啻于一剂良药。现实生活中还有少数人,尤其是某些青年人,身体有病却不听医生劝告,不注意调养身体,不愿接受治疗,或者虽然接受治疗也是心不在焉,因而影响疗效。对于这种情况,医生应该用严厉的言语,甚至恐吓的言语,使之提高警惕和接受治疗,这种诈术也是允许的。《黄帝内经》有一段记载:一个病人接受针刺治疗时态度不认真,因而气血散乱,这样会降低疗效。为了改变其精神状态,医生拿出一根长针给病人看,并且说:"我就要深刺下去了。"不过,真的刺入深度,还是根据病情决定的。

## 29、气失则师散,神衰则体病

心理学作为一个独立的学科不过一百多年的历史。但是,自从1879年德国科学家冯特创立第一个心理实验室以来,心理学的研究象雨后春笋,在人文科学领域得到迅速而广泛的发展。军事心理学和医学心理学就是心理学的两个分支学科。

心理学作为一个独立学科虽然是近代的事,但是对军事、医学等不同领域里人们的心理现象的研究却有着相当久远的历史。古代兵法家早就阐述过:对于敌人,是使之"气失而师散,虽形全而不为之用"(《尉缭子·战威》),那么对于自己军队呢?则应同仇敌忾,鼓舞士气,具有英勇顽强、前赴后继的斗志。通俗地说,灭敌人的威风,长自己的斗志,就是心理战的两个基本方面。

"长自己的斗志",激发我军士气这一军事上的心理战理论,早已被移植到医学上来。

《黄帝内经》早就注意到这一点,指出:"人之情,莫不恶死而乐生,告之以其败,语之以其善,导之以其所便,开之以其所苦,虽有无道之人,恶有不听者乎。"它包括四个方面的意思。

第一,"告之以其败",就是向患者指出疾病的性质、产生疾病的原因、疾病的危害、病情的轻重深浅,引起病人对疾病的注意,使病人对疾病具有认真对待的态度,既不轻视忽略,也不畏惧恐慌。

第二,"语之以其善",就是要耐心地告诉患者,只要及时治疗,积极与医生合作,按照医嘱进药或针灸,是能够恢复健康的,以增强病人战胜疾病的信心。

第三,"导之以其所便",就是要告诉患者如何进行调养,知道养生的方法,能自我进行调理养病。

第四,"开之以其所苦",就是要帮助患者解除紧张、恐惧、消极的心理状态。

不难看出,其基本思想是医生通过耐心细致地说明道理及有关情况,引导患者从心理上"杂于利害",使之既看到有助于治疗的有利条件和希望,又正视妨碍治疗的不利条件及可能产生的恶果,从而积极主动地配合医生治疗,趋利避害,尽早祛除疾病。在中医学史上,它被认为是说理开导治疗法的源头。

清人龙启瑞《涵劳楼古今文钞·病说》中就有一个用说理开导疗法治愈忧郁病人的故事。

龙启瑞去看望一个平时吃得很多而又长期腹泻的病人。病人自患腹泻病后,医治三个月也不见效,于是每天呆坐不动,以免病情加重,或仰卧于床,专思治病,然而病情反而一天天加重。龙启瑞诊断后说:"你所患腹疾,只要调整饮食,避免乍寒乍热,到时自然会好。倒是你整日忧心忡忡,不事劳作,精神就会出毛病。神是人体之主,神衰则体病。劳苦的人睡在悬崖绝壁处而不会下跌的原因,在于他能全神。一个小孩遇到老虎不但不怕,反而拿起棍棒追打,是由于他不知道老虎会伤人。你其实并无大病,却整天愁眉不展,疑神疑鬼,这才是你的病根。只要做到无忧无虑,怡然自得,病就会好的。"病人觉得言之有理,便依此而行,不出三日,果真腹泻终止,精神复原。

现代医学认为,对于那些已经知道病情的癌症患者应注意运用说理开导的方法进行心理治疗。首先,医生必须以同情、关心和亲切的态度耐心认真地倾听患者对病情、心理体验和要求的叙述,切忌轻易武断的否定、打断话题、过早解释劝说或表现出不耐烦情绪,以免引起病人的不信任而导致治疗失败。这样不仅有利于了解患者的心理矛盾和个性特点,为以后治疗打下基础,同时,通过引导患者自由表述,发泄出内心积聚很久的压抑情绪,可以达到释放心理能量和缓解症状的目的。在此基础之上,医生再有的放矢地针对患者的病症和病程预后,给予科学务实的说明、解释和保证。但是重点要讲清光明的前景,帮助患者树立顽强与疾病作斗争的信心,消除任何影响康复的不良情绪。反复向患者说明疾病可以治愈,保证身体可以康复,以及各种有效的治疗措施,有利于患者形成良好的心理支持,增强自我心理防卫功能,促进疾病康复或向良性方面转化。这对情绪强烈焦虑不安的患者,尤其是有消极自杀观念的患者特别重要。

《黄帝内经》说得好,人都是乐生恶死的。如果医生具有"杂于利害"的心理素质,并将之传递给病人,使之放弃心理包袱,看到康复的希望,那么对于医生和患者互相配合治疗,争取良好疗效,无疑是有百益而无一害的。

## 30、用乡导得地利,依归经利药效

在现代军事技术条件下,军队借助于各种通信导航设备及十分详细的地图,即使远行万里也不至于迷失方向。古代军队则不然,在既生疏又复杂的地区长途行军,由于难以事先了解前进方向的地形情况,他们通常要请一些熟悉该地区山川地理的人做向导。

孙子断然指出:"不用乡导者,不能得地利。"天时、地利、人和是军事上取得胜利的三大要素。这三大要素在其它领域只是表现形式不同,同样也是重要因素。任何领域,三者俱备则大事可成。

这三大要素,相互关联,缺一不可。如果在军事上一旦不能得地利,便很可能对战略全局造成不利的影响,甚至导致全军覆没。由此可见行军作战中若能有乡导引路,无疑有助于部队始终保持正确的方向,在有利的地形条件上充分保持和发挥战斗力。

清代医学家徐大椿认为用药治病决不能盲目从事,也应当象行军作战那样以向导引路。他在《用药如用兵论》中清楚地指出:"辨经络而无泛用之药,此之谓乡导之师。"

"辨经络",意为按经选药。古代医学家们在长期的实践中发现,每种药物在人体所起的作用都有一定的适应范围。如同属寒性药,虽然都具有清热作用,但有的偏清肺热,有的偏清肝热,有的偏清胃热;再如同是补药,有的是补肺,有的是补脾,有的是补肾。因此,中医药理学根据脏腑经络学说,结合药物的作用,把所有药物分别归于十二经,以说明某药对某些脏腑经络的治疗作用,并要求按经选药,这便形成了药物的归经理论。归,即归属;经,即经络及所属脏腑。

因为辨经络如同向导,一旦发生错误便会用错药,那就如同把军队引入误区,有全军覆没的危险,因此这里似乎有必要略花一点笔墨加以介绍。

经络之说完全是国粹,为中国医学所独创。它乃是运行全身气血、联络脏腑肢节、沟通人体上下表里内外的通路。它有经脉与络脉之分。经脉是主干,似径路;络脉是分支,似网络。经脉大多循行于人体深部,络脉则循行于较浅的部位,有的则显现于体表。经脉有一定的循行径路,而络脉则纵横交错、网络全身,把人体所有的脏腑、器官、孔窍及皮肉筋骨联结成统一的有机整体。络脉都直接或间接地与经脉相连,构成完整的经络系统。

在这个有机整体中,最主要的是十二经脉,它是运行气血的主要通道。这十二经脉对称地分布在人体的前后及两侧,循行于上肢或下肢的内侧或外侧,每一经分属一脏或一腑,故十二经脉中每一经脉的名称包括手或足、阴或阳、脏与腑三个部分。手经行于上肢而入脏腑,足经行于下肢而入脏腑;阴经行于四肢内侧属脏,阳经行于四肢外侧属腑。各经脉的名称是:

```
        ┌—手太阴肺经
手三阴经 ┼—手厥阴心包经
        └—手少阴心经
        ┌—手阳明大肠经
手三阳经 ┼—手少阳三焦经
        └—手太阳小肠经
        ┌—足太阴脾经
足三阴经 ┼—足厥阴肝经
        └—足少阴肾经
        ┌—足阳明胃经
足三阳经 ┼—足少阳胆经
        └—足太阳膀胱经
```

因为经络能沟通人体内外表里,在病变时,体表的疾病可以影响到内脏;内脏的病变也可以反应到体表。因此,人体各部分发生病变时所出现的证候,便可通过经络获得系统的认识。

如肺经病变,每见咳嗽、气喘等症;肝经病变,每见胁痛、抽搐等症;心经病变,每见心悸、失眠等症。将药物的功能与脏腑经络密切结合起来,就可以说明某药对某脏腑经络病变起着主要治疗作用。如桔梗、杏仁能治咳喘胸闷,故归肺经;羚羊角、钩藤能熄风止搐,故归肝经;朱砂、茯苓能安神定悸,故归心经等。

各种药物归经的范围并不是完全一致的,有的药物可归一经,有的药物可归数经,这说明它们或对某一经有治疗作用,或对某几经有治疗作用。如石膏归肺胃二经,是说明石膏既能清肺热,也能清胃热;党参归脾肺二经,是说明党参既能补脾气,也能补肺气。

与一药可兼入数经一样,由于脏腑经络病变具有传变性,可以互相影响,所以某一脏腑经络的病变有时亦需要同时应用数经之药。如肺病而见脾虚症,除用治肺药外,还可用补脾的药物,因脾能健运,可以促进肺病痊愈。可见,临床应用归经理论时切忌不要简单对号入座,而要根据脏腑经络之间的关系及病变部位而灵活运用。

如果掌握了药物的归经,治疗时按经选药,便如同行军作战时有了好的向导,有利于避免盲目性——充分发挥药物的性能,提高治疗效果。

例如:肝热目赤,在清热药物中,就可选用能清肝热的药物治疗;胃寒腹痛,在祛寒药物中,就可选用温胃止痛的药物治疗。所以,在重点熟悉药物功用的同时,进一步明确和了解某些主要药物的基本归经是十分必要的。

医生掌握药物归经的特点之后，不仅可以根据疾病的部位属于何脏何经，而使用相应的药物组成方剂，同时还知在中医大多数成方中都须选用一味向导药，即方剂配伍君臣佐使中的"使药"。使药又叫引经报使，它有将整个方剂的药物作用引向特定部位的功能。

如桔梗引诸药入肺，肉桂引诸药入肾，川牛膝引诸药行下肢，桑枝引诸药行上肢，升麻引诸药上升，沉香引诸药下降等，都是临床经常的选药方法。

## 31、擒贼擒王,治病治根

"射人先射马,擒贼先擒王。"这是唐代大诗人杜甫《前出塞》中的名句。素有"民间兵法"之称的《三十六计》中,"擒贼擒王"即为第十八计,并解释说:"摧其坚,夺其魁,以解其体。龙战于野,其道穷也。"大意是,摧毁敌人的主力,抓住它的首领,就可以瓦解它的整体。好比龙出大海来到陆地上,什么办法也就没有了。

打敌要害,从来都是兵家要诀。诸如《孙子兵法》所说的"避实击虚"、"先其所爱";《孙膑兵法》所说的"批亢捣虚",都是这个意思。所谓"虚"、"爱"、"亢"之类,即指敌人的指挥部、后勤基地、防御重心等。我们不妨举一个古代的典型战例来说明这个原则。唐朝李愬雪夜袭蔡州的战例堪称这一原则的最好注释。

**唐宪宗派兵攻打占据蔡州**(州治在今河南汝南)自立为王的吴元济,但进攻不力,战争呈胶着状态。公元816年冬,将军李愬得知蔡州兵力空虚后,决心直捣敌人的大本营。为确有把握,他先发兵攻打与蔡州紧邻的吴房,歼敌千余人即撤兵回营,既试探了敌人的城防力量,又避免了敌人的震恐。十月十五日夜晚,北风怒吼,大雪漫天,李愬亲率五千精兵直奔蔡州城。蔡州虽为吴军大本营,但三十多年没有战事,又逢风雪天气,故而戒备松弛。凌晨之时,唐军赶到蔡州城下,并迅速攀登城墙。吴元济从梦中惊醒,仓惶组织抵抗。怎奈唐军已四面入城,势不可挡,吴元济只得束手就擒,整个淮西战场的吴元济军随之而相继投降。

清代医学家徐大椿认为,用兵作战要讲究捣敌之穴,用药治病也要讲究捣敌之穴。所以,他在《医学源流论》中以用兵比喻治病时说:"病方衰,则必穷其所之,更益精锐,所以捣其穴。"他的意思是说,用药治病必须象打仗那样跟踪追击,扩张战果,集中精兵锐卒,捣毁敌人老巢。

治病治根,或说治病治本,是中医一系列治疗原则中最基本的一条。所谓"本"是相对于"标"而言的。任何疾病的发生、发展过程,都存在着主要矛盾和次要矛盾。"本"即是指病变的主要矛盾和矛盾的主要方面,起着主导的决定的作用;"标"是病变的次要矛盾和矛盾的次要方面,处于次要和从属的地位。因此,标本是两个相对的概念,可用以说明病变过程中各种病症矛盾双方的主次关系。如从正邪关系来说,正气是本,邪气是标;从病因与症状来说,病因是本,症状是标;从病变部位来说,内脏疾病是本,体表疾病是标;从疾病先后来说,旧病是本,新病是标;从疾病的传变来说,原发病是本,续发病是标,等等。

识别标本,才能进而医治标本。中医治病有一段流传久远的要诀:"见痰休治痰,见血休治血,无汗不发汗,有热莫攻热,喘生休耗气,遗精不涩泄,明得个中趣,方是医中杰,行医不识气,治法从何据,堪笑道中人,未到知音处。"

这段要诀说明,任何疾病的发生、发展总是要通过若干症状显示出来的,但这些症状只是疾病的现象,还不是疾病的本质。只有充分收集、了解疾病的各方面情况,通过综合分析,透过现象看本质,找出疾病的根本原因,从而选用适当的治疗方法,从根本上治疗,方能彻底祛除病根。

例如病人痰多,不一定用涤痰之法,因为"脾为生痰之源",健脾亦可使痰不再生;出血症不一定要用止血之法,因为"气为血帅",补气可摄血则出血自止;无汗不一定用发汗药,因为"汗为心液",可以用补养心液的方法治疗;发热的症状不一定用清热法治疗,因为"阴虚生内热",补阴其热自退;气喘的病人不一定服宣肺药,因为"肾主纳气",可以用补肾方法使气沉丹田,其喘自止;男子遗精病不一定用补肾涩精法,也不一定用泻火止遗法,因为肾"受五脏六腑之精而藏之",只要找出有病的脏腑,予以恰当治疗,遗精病就会痊愈。当然,上述"休治"、"莫攻"并非绝对不治、不攻,而是要求辨清疾病的根本证候,再予治疗,而不可头痛医头,脚痛医脚,只治标不治本,或只顾局部不顾整体。

治病治本包含若干具体的原则,一般说来,皆当治本,或标本兼治。少数标病危急者,也可先治标病,后治本病。运用之妙,就要依靠医生慧心独用了。《曹颖甫先生医案》中有一则病例就是说明灵活运用标本兼治原则的。

曹颖甫是近代一位颇有名气的医生。他曾治愈过一位疝痛患者。患者求治时,睾丸左大右小,小腹左旁有一肿包,大如小馍。饮食时,先是腹部鼓胀,不久即大痛,经一小时后,自觉病处有咕咕之声,痛乃逐渐减弱。患者曾经四处求医,先后经数十位医生诊治皆无明显效果,几乎痛不欲生,只愿求死而已。曹医生诊断后,认为此病起因在于寒湿内侵,留滞厥阴肝经,气血郁滞而致。根据"治疝皆归肝经"、"治疝必先治气"的原则,故当暖肝温肾,行气止痛。于是,他采用一系列药物,温补肝肾以治其本,行气逐寒以治其标,不久便使患者病体康复。

## 32、老敌之师,扶正祛邪

俗话说:"病来如山倒,病去如抽丝。"有些严重的疾病初发之时,的确来势凶猛,难以阻挡。在这种情况下,如果在治疗上一味"贪速",投之以猛药,不但无益于治疗,反而还会刺激病情加剧。所以,清代医学家徐大椿认为,正确的方法应当是"老敌之师"。

"老敌之师",原是中国古代兵法中的一条原则。所谓"老",表示"使……衰弱"、"使……疲怠"。显然,"老敌之师",意在使敌人的军队衰弱、疲怠,失去战斗力。

春秋末期著名谋略家范蠡有一个观点:"尽敌阳节,盈吾阴节而夺之"。他认为,聪明的将领必须善于设法消耗敌人的力量,同时隐蔽地发展自己的力量,待敌军衰弱疲怠,我军气壮力足时,再一战而胜。徐大椿正是在这个意义上提出治病也当"老敌之师"。他认为,"病方进,则不治其太甚,固守元气,所以老其师"(《用药如用兵论》)。不难看出,对于来势凶猛的疾病,徐氏并不是主张放弃治疗,而是要求在加强患者抗病能力的同时,循序渐进,适当治疗。用中医的行话来说,就是要扶正祛邪。

中医认为,疾病的过程,在一定意义上可以说是人体的抗病能力(正气)与致病因素(邪气)矛盾双方相互斗争的过程。邪胜于正则病进,正胜于邪则病退。因而治疗疾病,就是要扶助正气,祛除邪气,改变邪正双方的力量对比,以利于疾病向痊愈方向转化。

**扶正**,就是使用扶助正气的药物或采用其它疗法,并配合适当的营养及功能锻炼,增强体质,提高机体抗病能力和自然修复能力,从而达到祛除邪气,恢复健康的目的。即所谓"正气内存邪不可干"、"正足邪自去"。此法主要适用于正气虚弱而邪气不盛,以正气虚为主要矛盾的病证。临床可根据病人的具体情况,分别运用益气、养血、滋阴、助阳等补法。

**祛邪**,就是使用攻逐邪气的药物,或运用针灸、手术等其它疗法,以祛除病邪,达到邪去正复的目的。即所谓"祛邪以扶正"、"邪去正自安"。此法主要适用于邪气盛而正气未衰,以邪实为主要矛盾的病证。临床可根据邪实的不同情况,分别运用发汗、攻下、消导、化瘀、涌吐、祛湿、祛风等治法。

虽然扶正与祛邪的具体方法不同,二者的目的却是完全一致的。因此,临床应用时必须彼此配合,根据具体病情,或先扶正后祛邪,或先祛邪后扶正,或同时并用。一般来说,先扶正后祛邪,适用于正气虚为主而邪气不盛之证;先祛邪后补正,适用于邪气盛为主而正气亦虚之证;扶正祛邪同时并用,则适用于正气虚兼邪气盛的病证。若单补正则助邪,若单祛邪则伤正,而补泻兼施,才能两全。总之,要能够"固守元气","老敌之师",扶正不留邪,祛邪不伤正。为具体理解这一原则的含义及其运用方法,我们不妨以元代名医朱丹溪治疗伤寒的一个病案为范例,略加分析。

据《古今名医案》记载:朱丹溪曾接治一位老年病人。老人因大冷天长时间在户外劳动,又冷又饿,患头痛恶寒发热、关节疼、无汗等症,并伴有神志不清,胡言乱语的现象。发病之初,老人自己服用了一些参苏饮,以发汗祛寒,但汗大出后却高烧不退。朱丹溪诊断后认为,老人患的是内伤证,因饥而胃虚,加以劳作、受寒,故致使寒邪侵入。于是,他以人参、黄芪、当归、白术、陈皮、甘草等药为主,并先后加减附子、芍药组成配方,让老人一天服用五剂,连服十日。病情稍有好转时,又让老人增加一些肉汁粥和苁蓉粥,辅以食补。十日之后,老人果然病去神安。

此案中,老人的病虽诱发于伤寒,而其矛盾的主要方面在于内伤。因年老体弱,虽有"头痛恶寒发热、关节疼、无汗"等典型的伤寒表实证,也不能攻表发汗,否则汗愈出正气愈虚。所以朱氏并不急于祛寒,而是先着手补虚,所开的药方中大多是补气养血之药,并先加附子助阳,后加芍药助阴,从而使胃气逐渐充实,肌体自然出汗,迫使体内寒邪不断疲怠,最终不得不随着汗水而逸出体外。

由此可见,老敌之师,扶正祛邪,不失为治疗邪盛正虚患者的良方,而且也是治病求本最基本的、最有效的途径。

## 33、断敌要道,阻隔病源

清代医学家徐大椿指出:"传经之邪,先夺其未至,则所以断敌要道也"(《医学源流论》)。在这里,他准确地用兵法语言表述了一个十分重要的医疗观点,那就是对于疾病要早治,而且要治其要害,防止扩散。

他所谓"传经之邪",指的就是通过经络侵入人体的病邪。伤寒学说认为,寒邪侵袭机体由肌表而入,可沿六经传变。六经即太阳、阳明、少阳、太阴、少阴、厥阴,原是人体十二经手足同名经脉的简称。汉代医学家张仲景以此分析外感热病在演变过程中所产生的各种证候,依据所侵犯的经络脏腑,病变的部位,受邪的轻重,邪正的盛衰,归纳成为六个不同的证候类型,称为六经病。六经病的发生、发展过程,实际上就是外感病邪由表入里、逐步深入的过程。太阳病是外感疾病初期,邪居表卫阶段;阳明病是病邪入里化热的极期阶段;少阳病为邪居半表半里之间的过渡阶段。邪气深入,正气已虚,则可传入三阴经而成三阴病。太阴病为脾虚湿盛阶段;少阴病为心肾阳衰阶段;厥阴病是寒热交错、阴阳胜复的病理阶段。

此外,温病学说也认为,病邪主要是从口鼻而入,其病变是按卫气营血或上、中、下三焦进行传变的。

大凡这种侵入人体的病邪,一般都遵循由外入内,由表及里,由浅入深,由轻到重的规律。因此《黄帝内经》早就指出,"邪风之至,疾如风雨。故善治者,治皮毛,其次治肌肤,其次治筋脉,其次治六腑,其次治五脏。治五脏者,半死半生也。"如不及时治疗,就会坐失良机,疾病日趋深重,恢复也就不易。

疾病沿经络而侵入腑脏，如同敌人沿道路而深入重地一样，具有明显的共同点。毫无疑义，军事上防敌入侵的原则应当是控制要点，在敌人必经之路（要道）上顽强抗击，务求歼敌，以免敌人深入重地。《孙子兵法》指出："能使敌人不得至者，害之也。"这里的"害"，意为妨碍、阻挠。用现在的话来说，就是要遏制敌人，使之不能到达战地。正如唐人杜佑注释此句时所言："出其所必趋，攻其所必救，能守其险害之要路，敌不得自至。"

那么，这一军事原则在医疗上如何体现呢？

元代医学家王好古曾经说过："衰热之法……譬如孙子之用兵，若在山谷，则塞渊泉；在水陆，则把渡口；在平川广野，当清野千里。塞渊泉者，刺俞穴；把渡口者，夺病发时前；清野千里者，如饥羸瘦弱，宜服大药以养正大"（《此事难知》）。此言以兵法比喻疗法，把徐大椿"断敌要道"的观点进一步形象化、具体化，使人们清楚地认识到，要想防止疾病传变，必须及时辨证施治，对症下药，灵活应用针灸、预防、药疗等各种方法，尽量把病邪遏止在未萌或初萌状态。

当然，"阻隔病源"的治疗法则并非王好古先生首创，《黄帝内经》中即已明确提出，并且具体应用于防治疾病的过程之中了。在掌握疾病发展规律的基础上，根据已经表现出来的某些轻微症状，推断出其必然的发展趋势，及时有效地采取阻隔措施，便能防止疾病加重而很快治愈，正如《素问·阴阳应象大论》所说："见微得过，用之不殆。"《素问·离合真邪论》描述这种治疗法则为"早遏其路"。现以五脏热病为例说明其具体运用。五脏热病在其将发未发之时，会在颜面部的特定位置上出现"赤色"，此时予以针刺治疗，即可防止热病发生。如"肝热病者，左颊先赤；心热病者，颜先赤；脾热病者，鼻先赤；肺热病者，右颊先热；肾热病者，颐先赤。病虽未发，见赤色者刺之，名曰治未病"（《素问·刺热篇》）。

## 34、守我岩疆，先安未病

中医理论认为，人的身体是一个统一的有机整体，其中最主要的是十二脏腑。《黄帝内经》中一段颇有意味的对话，生动形象又简明扼要地介绍了这十二脏腑的功能及相互影响、相互作用的特点。

黄帝问医官岐伯："我希望听你讲一下十二脏腑在人体内的相互作用，有无主从的区别？"

岐伯为了既清楚又形象地说明，决定以国家官制来比喻各个脏腑。他解释说："在人体内，心的重要性就好比君主，人们的聪明智慧都是从心生出来的。肺好象是宰相，主一身之气，人体内外上下的活动，都需要它来调节。肝譬如将军，谋虑是从它那儿来的。胆是清虚的脏腑，具有决断的能力。膻中象个内臣，君主的喜乐，都由它透露。脾胃受纳水谷，好象仓库，五味化作人体的营养，是由它那儿产生的。大肠主管输送，食物的消化、吸收、排泄过程是在它那儿最后完成的。小肠的功能，是接受脾胃已消化的食物后，进一步起到分化作用。肾是精力的源泉，能产生出智慧和技巧来。三焦主疏通水液，周身行水的道路，是由它管理。膀胱是水液聚会的地方，经过气化作用，才能把尿排出体外。以上十二脏腑的作用，不能失去协调。当然，君是最主要的。它如果得力，下边就能相安。这是根本的道理。如果依据这个道理来养生，就能长寿，终身不致有严重的疾病。如果根据这个道理来治理天下，国家就会非常昌盛。反之，如果君主不得力，那么，十二官就成问题了。而各个脏腑的活动一旦失去联系，形体就会受到伤害。对于养生来说,这样是很不好的。对于治国来说,这样做,国家就会有败亡的危险,实在值得警惕呀!"

两千多年前,中国古代医学家能对人体内部各个器官的机能作出这样具体的分析和描绘,今天读来令人慨叹不已。岐伯的解释清楚地说明,十二脏腑不仅各有其职,而且彼此相连,互相影响。倘若其中某一脏腑产生病变,则必然产生向其它脏腑幅射的趋势;相反,如果及时增强其它脏腑的抗病能力,也必然有利于遏制病邪的发展和蔓延。

由于人体组织器官具有这种特点,历代良医无不重视《黄帝内经》中"治未病"的观点。"治未病"包含两层意思,既指防治肌体未受病邪之前,又指在病邪侵入之后及时保护未受侵袭的脏腑,因此它是预防和治疗通用的原则。

> 托名扁鹊的《难经》从治疗学上阐发了这一原则,提出:"上工治未病,中工治已病者,何谓也?所谓治未病者,见肝之病,则知当传之于脾,故先实其脾气,无令得受肝之邪,则知肝当传于之于脾,故曰治未病焉。中工者,见肝之病,不晓相传,但一心治肝,故曰治已病也。"

这是一个相当精辟的见解,鲜明地阐述了既病防变的观点。从他所举的这个治肝实脾的病例可以看出,医生高明不高明,在这个问题上就是一个分水岭。高明者,便能在治疗病变脏腑的同时,及时采用具有健运滋补性质的药物增强相关脏腑的抗病能力,遏制病邪进一步传变。

> 清代医学家叶天士,根据温热病伤及胃阴之后,病势常进一步发展,往往耗及肾阴的特点,主张在甘寒养胃的方药中,加入一些咸寒滋肾的药物,并提出了"务在先安

未受邪之地"的防治原则。

清代医学家徐大椿在《医学源流论》中提出:"横暴之疾而急保其未病,则所以守我之岩疆也。""岩疆"者,即专指那些地势险要,地位重要之地。那么"守我岩疆",就是要严密防守那些很可能遭受病敌侵袭的部位。

徐大椿将这一法则引入治病领域,确实独具慧眼。其意在指导人们,对于传变性较强的疾病,要象军事上固守要塞以拦截来犯之敌那样,在积极治疗的同时,迅速保护与病变部位密切相关的脏腑,从而扶持正气,驱逐邪气。

## 35、焚敌资粮,断除病源

"兵马未动,粮草先行。"这是古人行军作战的经验之谈,《孙子兵法》曾明确指出:"军无辎重则亡,无粮食则亡,无委积则亡。"辎重,即车辆及各种装备;委积,即各种物资储备。战争中如果没有辎重、粮食、委积等充足的后勤保障,军队就必然处于灭亡的境地。于是,搞好运输,畅通粮道就成为后勤供应的重要任务。基于这种认识,孙子认为聪明的将领应当善于"断敌粮道"。断敌粮道的最好办法莫过于火攻。因而,他在论述火攻用途时着重强调"火积"、"火辎"、"火库"、"火队"("队"通"隧",道路),即烧毁敌人的军需、辎重、仓库、粮道,使敌人失去继续战斗的条件。

无妨举一个典型战例来说明一下吧。三国时期有场著名的战争,那就是官渡之战。

公元200年,曹操与袁绍对峙于官渡。袁军有数十万之众,曹军却只有数万人,形势明显有利于袁绍。但是,在万分危急的关头,曹操得知袁绍刚刚在官渡附近的乌巢建立一个后勤供应基地,以保障袁军的粮食和草料。于是,他一面留下主力继续抵抗袁军的攻势,一面亲自率领五千精兵偷袭乌巢。曹操让所有精兵装扮成袁军,衔枚勒马,星夜奔驰,趁乌巢守军尚在熟睡之时,一把大火将数十万担粮食化为灰烬。袁军失去粮食供应,顿时上下一片惊慌,不战自乱。曹操乘机反攻,大获全胜。所以,古代兵法指出:"敌既无粮,其兵必走,击之则胜"(《百战奇法》)。

清代医学家徐大椿可谓是善于创造思维的学者,他认为治疗某些疾病也可以采用"焚敌资粮"的办法。他在《医学源流论》中提出:"挟宿食而病者,先除其食,则敌之资粮已焚。"认为治疗因吃得过饱,引起积食的疾病,应当象兵法上焚敌粮草那样,先清除产生疾病的根源。"先去其食"就是用吐、泻等方法排除胃里的宿食或者致病食物。元代名医罗天益治疗一位因肉食不当而致病的病人就是生动的史例。

有一壮年男子与朋友一起打猎,猎得野兔几只,于野外生火烧烤,朋友们每人吃一块,他却比别人多吃一块。傍晚回到住地后,因觉得口渴,又喝了几碗奶汁。当天晚上,这位男子腹胀如鼓,疼痛难忍,卧而欲起,起而又卧,吐不出,泻不下,手足无所措,他的家人急忙请来罗天益诊治。罗氏诊断之后,认为病人系因暴饮暴食而损伤肠胃。烧肉干燥,多食则引起口渴,干肉受奶汁稀释,胀满肠胃,导致腹胀疼痛,起卧不安,手足无措。于是,他马上让病人服用具有下泻作用的备急丸和具有催吐作用的无忧散,分别排除肠胃中的积食。三天之后,再以稀粥及具有益气作用的参术之药调养。不出七天,病人便痊愈了。

徐大椿虽然只提到"挟宿食而病者"应采用"焚敌资粮"的办法,但是,我们完全可以由此举一反三,从中揭示出带规律性的道理来,那就是各种疾病都应首先清除病源,以利治疗。无论是外感六淫、内伤七情还是饮食失调、房事失节、劳逸失度、外伤失防所引起的疾病,都应当先"焚敌资粮",清除病源,否则越治越忙。下面这个例子就充分说明了这一点。

独生女娟娟,在父母、爷爷和奶奶的百般爱护下长到了十五岁,可她那纤弱的身体一受风吹全身就起鸡皮疙瘩、头痛、打喷嚏、流清鼻涕、还怕冷、低烧、全身疲乏无力,动一动就气短、自汗,稍不注意就感冒。一年四季,反复无常。

这可把她家里人急坏了,常常四处求医,八方打听。有的医生说,她病在皮毛,俗称"伤风"、"着凉","小毛病不要紧,给她开个处方,吃了包好";有的医生则说,她的病一不要打针,二不要吃药,经过七天七夜就自然会好。但是,几年过去了,医来医去总是收效甚微。

一天,她姑爷爷来家,一家人不免又谈到了娟娟的病。这位年逾花甲,行医几十年的老中医看了看娟娟的气色,又环视一圈居室,说:"你们把门窗关得这么紧,生怕漏一点风进来。阳春三月了,还让她穿那么一件厚棉衣。娟娟好比温室里的花朵,见不到阳光,经不起风吹雨淋,所以经常感冒。"他接着分析:"娟娟的病概括起来就是一个虚字,也就是正气虚,由正气虚抵抗力下降而导致细菌、病毒便乘虚而入。"于是,老人用黄芪、白术、防风三味组成药方,以帮助娟娟益气健脾,解表祛寒。同时,他强调,在体质虚弱的时候,最重要的是坚持体育锻炼,多做户外活动,平时要到大自然中去经风雨,见阳光,增强体质,使正气充盈,才不为风邪所侵袭。不言而喻,老人这一番"焚敌资粮",杜绝病源的良方终于治好了娟娟的病。

## 36、断敌内应,标本兼治

生活中常有这样的事,"福不双至,祸不单行。"本有老年性慢性支气管炎病又患感冒,或者本有脾胃慢性病又患肺炎。在中医理论中,原发病称作"本",续发病称作"标"。清代医学家徐大椿用兵法的眼光观察医疗,形象地指出,原发病好似敌人的内应,续发病则好似正面来犯之敌。内外夹攻,腹背受敌,如何迎敌呢?徐大椿断然指出,必须采取"断敌内应"的治疗方法。他说:"合旧疾而发者,必防其并,则敌之内应既绝"(《医学源流论》)。在这里,徐氏所说的"内应"指的就是原发病,也就是他所说的"旧疾"。不让旧疾复发,就如同断绝了新疾的内应。

徐大椿确实懂得兵法。从军事上说,敌之内应是个别的、少量的策应主力部队的敌人。战争中,大凡遇到敌人主力与小股内应同时威胁我军防御时,正确的打法是,以我之主力歼灭正面来犯之敌,对于敌之内应只须以少量兵力或钳制或阻绝,不使骚扰即达目的。徐大椿将这一军事原则非常生动地运用到了医疗上,这是对传统医学标本兼治理论的正确发挥。

标本兼治,既可先治标后治本,也可先治本后治标,还可以既治标又治本。此理论除适用于同时治疗新旧疾病外,还可以从因果、表里、主次等方面去理解和运用。如外感热病,热邪入里,由于里证实热不解而阴液大伤,表现为腹满硬痛,大便燥结,浑身发热,口干唇裂,舌苔焦燥等正虚邪实、标本俱急的证候。对于这种病证就应当标本兼治,泻下与滋阴两法同用,即清泻实热以治本,滋阴增液以治标。若仅用泻下之法,则有进一步耗伤津液的可能,而单用滋阴之法,则又不足以泻下在里之实热。如果两法兼用,则泻下实热即可存阴,滋阴润燥有利于通下,如此标本同治,相辅相成,便可达到邪去液复的目的。又如痢疾,可见腹痛,里急后重,泻下赤白脓血,舌苔黄腻,脉象滑数。其病因湿热为本,故其治疗应以清热利湿之法以治本,还应配合宽肠理气的方法,以解决腹痛、里急后重之急,这也是标本兼治法的具体体现。

疾病的变化是错综复杂的,标病与本病不是固定不变的,它们可以在一定条件下相互转化,因而标本兼治应注意根据病情发展而灵活掌握,而不是不分主次地平均对待。或治本病为主,兼治标病;或治标病为主,兼治本病。如肺痨病,肺阴亏虚为本,潮热、盗汗、咳嗽等症状为标,治疗时应当以治本为主,兼以治标。但是,如果大量咳血,又当以治标为主,兼以治本。

断敌内应,防敌合并,首先必须弄清谁是内应,谁是外敌。标本兼治也是如此,辨证施治过程中必须分清哪是标病,哪是本病,才有可能抓住主要矛盾,及时治疗,以避免"屋漏更遭连夜雨,船迟又遇打头风。"

## 37、地有六形,医有六经

清代医学家柯琴在《六经正义》一书中发表了一通高论,颇能给人启迪。他说:"兵法之要,要明地形。必先明六经之路,才知贼寇所从来,知某方是某府的来路,某方是某府的去路。来路是边关,三阳是也;去路是内境,三阴是也。六经来路各不同,太阳是大路,少阳是僻路,阳明是直路,太阴近路也,少阴后路也,厥阴斜路也。客邪多从三阳来,正邪多由三阴起,犹外寇从边关至,乱民自内地生也。明六经地形,始得握百病之枢机,详六经来路,乃得操治病之规则。"

战争是在一定的空间中进行的,山林、沼泽、平原、戈壁等都是用武之地。所以,《孙子兵法》十分重视对地形的认识和应用,提出"夫地形者,兵之助也"、"知天知地"等名言。可贵的是,孙子还对军事地理进行了分类,比如,把战略地理分为九种:散地、轻地、争地、交地、衢地、重地、圮地、围地、死地;把战术地形分为六种:通地、挂地、支地、隘地、险地、远地。

柯琴从军事地理在战争中的作用上受到启发,加以联想,提出了认识人体"六经"如同战争中识别地形的观点。

六经辨证就是根据人体抵抗力的强弱,病势的进退、缓急等,将外感疾病演变过程中出现的证候进行分析,综合为太阳、阳明、少阳、太阴、少阴、厥阴六经病证,以此来归纳证候特点、病变部位、寒热趋向与邪正盛衰,最终作为诊断、治疗的依据。按照柯氏的解释,六经病证统属阴阳两大类,其中三阳病证多来自外邪入侵所致,病在肌肤,三阴病证则多由内邪作怪所致,病在五脏。三阳病既由外邪侵入引起,所以当用驱邪之药为主,如麻黄汤发散太阳之寒邪,承气汤攻逐阳明之实热,小柴胡汤和解少阳之郁火;三阴病既由"内乱"为前提,因而治疗时应当重视安抚正气,如理中丸补脾阳而去寒以治太阴,真武汤壮肾阳而去水以治水阴,乌梅丸寒热并投以治厥阴之病。

## 38、治齿如治军,治目如治民

古代兵学家曾经指出:"军容不入国,国容不入军"(《司马法》)。这就是说,治军的原则和方法不能用来治理民众,同样治理民众的原则和方法也不能拿去治理军队。为什么呢?军队是武装集团,令行禁止,赏罚严明,管理十分严格,显然不能用军队的规章制度去要求老百姓。

有趣的是,宋朝大诗人苏东坡在记述朋友张文潜对他的告诫时,却把"军容"移用到"医容"上来了。苏东坡写道:有一天他同张文潜几个人出游,他目疾犯了,多次用热水去洗目。这时张文潜告诫他:"目有病,当存之,齿有病,当劳之,不可同也。治目当如治民,治齿当如治军"(《东坡志林》)。存,休养之意。劳,运动之意。

张文潜的这番话,显然是一个妙喻。确实,经常叩齿,便是行之有效的坚齿方法。即使已患齿疾,叩齿仍然有效。北齐颜之推在《颜氏家训》中写道:"吾尝患齿,摇动欲落,饮食热冷,皆苦疼痛。见《抱扑子》牢齿之法,早朝叩齿三百下为良。行之数日,即便平愈。今恒持之。"

人的眼睛是个非常娇嫩的器官,其构造十分精致。为防止受到伤害,它有一套周密的防护装置。人们常常形容爱惜一件东西如同爱护自己的眼睛一样。俗话说,"眼睛里容不下一粒沙子",说明即使一颗微小的沙子弄进眼里,也会痛苦难忍,轻者把眼角膜上皮擦伤,重者可造成球内异物或眼球裂伤。所以保护眼睛最重要的方法之一,就是使它不要过度疲劳,如同国家对待民众那样,尽量减少他们的负担而促进生产发展,以休养生息。尤其是当眼睛有病时,更不适宜再用热水去刺激它,俗话说"眼疼不用吃药,全凭睡觉",就是指给眼睛以充分休息,有助于其病恢复。

眼病不宜用热水洗,牙病却可常"敲打"(叩齿),恰如治民与治军不同。

## 39、以寡胜众,同病异治

读了《医学源流论》关于"一病分治"的论述,真是如饮佳酿玉醇,既可看出徐大椿其人高深的医学功底,也可看出他对于《孙子兵法》造诣很深,否则,是很难提出如此众多的真知灼见的。

就拿"一病分治"来说,徐大椿居然能用"以寡击众"的兵法原理去解说,去运用,并且真正做到融会贯通,真是令人钦羡不已。

我们知道,众寡,是中国古代兵学中有关军事实力的一对基本范畴。众,表示兵力多;寡,表示兵力少。就一般规律来说,通常是以众胜寡,但是如果灵活运用兵力,以寡胜众也是可能的。《孙子兵法·虚实篇》指出:"形人而我无形,则我专而敌分;我专为一,敌分为十,是以十攻其一也。"孙子紧接着说明:"则我众而敌寡,能以众而击寡者,则吾之所与战者,约矣。"意思是说,当我兵力弱小时,先用假象隐蔽自己,迷惑敌人,分化敌人。这样虽然从总体上说,我寡敌众,我一敌十;但是由于我兵力集中,因此对于当面之敌来说,我是以十击一,以众击寡。

以寡击众,也就是以少击多,以弱击强。取胜的方法说起来五花八门,多种多样,但是核心的一条就是造成敌人兵力分散,我的兵力集中。这一条作战原则古今中外概莫能外。令人惊异的是,徐大椿竟能将此原则活用到医疗上。他在《医学源流论》中指出:"一病而分治之,则用寡可以胜众,使前后不相救,而势自衰。"

所谓"一病而分治之",是说同一疾病出现不同症状,或者在不同阶段反映的性质不同时,就应当采用不同的方法治疗。所谓"用寡可以胜众,使前后不相救",喻指治疗用药不多但收效甚高,并使多种疾病之间无法相互传变,如同溃败之军"前后不相救"。

例如痢疾一病,有便脓血、里急后重、腹痛等症状,应以行气活血分别治之。行气则里急后重自除,腹痛亦随之而止,活血则便脓血自愈。

又如麻疹一病,在病变的不同发展阶段中可能出现不同的症状,治疗也就应当有所区别。麻疹初期病邪在表,宜发表透疹;中期多为肺热壅盛,宜清热解毒;后期多是余热未尽,肺胃阴伤,宜养阴清热。如此分而治之,便于各个击破,缓解病情,使病邪不能进一步传变,以致逐渐衰退。这实际上就是中医理论上常说的同病异治之法。

"一病而分治之",或者说同病异治,还有一层含义,那就是,相同的疾病,由于病邪性质、人体反应及病人生活环境等因素的不同,可以用不同的方法治疗。

疾病如敌。例如哮喘此"敌"可分为寒喘、热喘、虚喘、实喘,治法(如同打仗的用兵方法)也就相应不同。每个人的体质强弱不同,性别、年龄、生活环境不同,同是哮喘,在老年人和壮年人之间,在体质强壮和体质虚弱的人之间会有不同的表现。即使同等年龄,同是体质虚弱的人,严寒的冬天与酷热的夏天,潮湿的霉季和干燥的秋天,也会使哮喘的类型各异。南方偏于温湿,北方偏于冷燥,用药的侧重也不同。所以即使在同一个人身上,哮喘用药也可由气候、地区、疲劳等因素而有所改变。

病情复杂，如同敌人众多，但是掌握了疾病发生演变的规律，就可因病用药，达到用力少而收功多的"以寡胜众"的奇效。

例如，有一个患哮喘二十年之久的病人，几年前在北方发作时非常怕冷，天稍转冷就容易引起哮喘发作，痰多如泡沫水样，经中医辨证认为是"寒喘"，经服以温性的小青龙汤药而见效。次年，此人到南方，路途疲劳引起发作，自己仍按小青龙汤配服，不但无效且气喘更厉害，面色苍白，说话无力，稍事活动就气短出汗。这是"气虚"表现，给以补气汤而制止了发作，两个月内没有复发作，体力也转强。但有一天闻到煤气与艾叶熏烟，又引起哮喘发作。当时已是秋天，气候干燥，病人仍深信补气汤，便自行配服，结果无效。医生诊视病人，只见面色发红，口唇干燥，舌色也红，并伴有头痛，显然已转为"热喘"。在给以凉性的麻杏石甘汤之后，才使哮喘平息下来。以后按哮喘的发生是与肾有密切关系的理论给以补肾治疗，使体质有了变化，以致几年不再发作。

## 40、捣敌中坚，异病同治

清代医学家徐大椿对于"同病异治"借鉴了《孙子兵法》的"以寡胜众"的原则，与此同时，他又对"异病同治"移用了《孙子兵法》的"捣敌中坚"的原则。他在《用药如用兵论》中写道："数病而合治之，则并力捣其中坚，使离散无所统，而众悉溃。"

所谓"数病而合治之，则并力捣其中坚"，是指病变的部位、脏腑的功能和发生的疾病虽然不同，但病机相同，就可抓住其共同点，采用相同的方法治疗。

例如虚寒泄泻、脱肛、子宫下垂同时病于一人，这些病症都是由于脾气虚所致，所以可以用补中益气的方药捣其中坚。一旦中气充盈，既可遏止各病症的发展势头，又可割断它们彼此的联系，从而使之"离散无所统，而众悉溃"。这种治疗方法，中医理论称之为"异病同治"。

从中医理论上来说，"数病而合治之"，或者说异病同治，还应当包含更广泛的意义。它不仅适用于数病齐发的情况，还适用于病异症同的情况。不同的疾病，中医辨证可以有同样的病机，可以有同样的辨证类型，治疗方法也就可以相同。

例如，患者徐某，因饮食不当，连续几天午后发热，口干体倦，小便呈红色并有滞涩之感，两腿酸痛无力，脉虚大，舌苔色淡，为肾阴虚，且肾火上升。本案应先从肾虚火旺治疗，即用六味地黄丸，可谓切中病机。又如,患者姚某,上吐下泻,没有任何食欲,见到食物就恶心,面色萎黄,神情困倦,从秋天到春天,经久不愈,属脾肾阳虚之证。脾虚不能运化水谷,所以引起恶心厌食;脾阳下陷,所以引起腹泻;饮食减少,致使气血不能生化,自然精神困倦,面色萎黄。医生用补中益气汤予以治疗,但减去其中的当归,加上肉果,以提升下陷的阳气,又用姜附补火以生土。四日之后止泻,但吐痰不减,再以八味丸与补中益气汤并用,很快消除痰源。

这两个病例,病证不同,病因相异,但表现为同样的病机,所以二者都可以用补中益气汤治疗,只不过酌情适当加减其中的一两味药而已。

实践证明,病症尽管千差万别,它们之间却有其共同之处。疾病毕竟都是通过人体反映出来的,它们之间就必然有着共同类型或共同规律,所以在不同疾病处于同一类型或同一阶段时,都可以用同一方法治疗。

例如疟疾、痢疾、结肠过敏、支气管哮喘、乳糜尿等,一旦出现精神不振、四肢无力、倦怠懒言、语音低怯、呼吸气短、动则多汗等一系列"气虚"的症状,就可采用补气的方药治疗。对于各个疾病来说,补中益气汤并不一定能治好这些病,因为它毕竟不是针对这些病的原始病因的,但有时在"气虚"的情况下,即使采用针对原始病因的治疗方法,例如疟疾采用杀灭疟原虫的药,痢疾采用杀灭痢疾杆菌的药,支气管哮喘采用解除支气管痉挛的药,也可能不完全见效。这时补气的方法因为扭转了"气虚"这一关键问题,从而改善了整体,加强了自身的抗病能力,从而靠人体正气杀灭了病菌原虫,解除了支气管痉挛;或者是由于使杀灭病菌原虫、解除支气管痉挛的药物得以发挥效力,病也就自然减轻或痊愈。这就是不同的疾病在其发展过程中,使人体变化处于一个共同的关键状态时,运用"异病同治"的方法,从而提高了疗效。

异病同治,并不是说抓住任何一个共同症状就可以施以相同治疗方法,而是要抓住具有关键意义的特征,以"捣其中坚",使对异病的治疗提高疗效。"胃口不佳"、"睡眠不安"等症状就很少拿来当作关键性特征对待,因为这些症状较少接触到疾病本质。而"肾虚"、"气虚"、"血瘀"等往往成为关键性特征,因为这些特征是比较根本性的,通常是疾病的根源,较多接触到疾病的本质。这就需要人们在辨证施治过程中仔细鉴别,准确抓住其"中坚"。

## 41、无击堂堂之阵，无刺熇熇之热

医家攻邪治病，犹如兵家治军打仗。医学经典《黄帝内经》说："兵法曰：无迎逢逢之气，无击堂堂之阵。刺法曰：无刺熇熇之热，无刺漉漉之汗，无刺浑浑之脉……方其盛也，勿敢毁伤，刺其已衰，事必大昌。"

军事理论告诫指挥官们，当敌军士气蓬（逢）勃旺盛、敌阵严整雄壮（堂堂）之时，不要逞匹夫之勇，冒然进击，这样做很容易吃败仗，即使侥幸获胜，也难免损兵折将，大伤元气。需等待时机，伺敌军士气衰落、阵容散漫之时，再以精锐之师奋勇击之，则己方只以最小的代价便可一鼓作气而取得全胜。人体患有疾病时，大多数情况下也含有敌与我两方面的势力。邪气即是敌方，使人疾病加剧；正气即是我方，是驱逐邪气恢复健康的内在因素。医生治病时，无论用针下药，还是按摩推拿，都是支援正气、驱除邪气的手段。而作为高明的医生，在施以各种治疗方法时，亦应选择有利时机，以保证取得最佳效果。这个时机就是病人的正气相对旺盛，致病的邪气相对虚弱的时候。

例如"医圣"张仲景用十枣汤治疗"水饮"病，就选择在人体阳气相对旺盛的"平旦"（清晨）时给药。现代实验研究也证实，"补阳药"早晨给药比晚上给药的效果强得多，这是由于早晨人体中的阳气正在上升，故此时给予支持收效显著。

选择正气相对旺盛、邪气相对衰弱之时施治,对于周期性发作或加剧的疾病尤其重要。例如疟疾当其发作时,邪气最为猖厥,正气最为虚弱,假若此时治疗,要想制服邪气必须使用剧烈的"虎狼之药",而虎狼药在制服邪气的同时,也必然损伤本来已经衰弱的正气,最后是敌我两败俱伤,所以医生不能轻易使用这种方法。而在疟疾未发作之前的休止期,邪气较弱,正气相对旺盛时治疗,则可使用小剂量或较平和的药物,这样既能驱除邪气,又最少地损伤正气,使疾病痊愈而很快康复。

所以《黄帝内经》说:"无刺熇熇之热,无刺漉漉之汗,无刺浑浑之脉。"熇熇,火邪炽盛。浑浑,邪盛而脉乱。漉漉,汗出如水是邪盛正衰之象。此时无论用针、用药,均非所宜。所以《黄帝内经·疟论》明确指出:"方其盛时必毁,因其衰也,事必大昌。"只有当邪气衰退之时进行治疗,才能收到事半功倍的效果。当然,对于那些病情危急,又非"间歇"性缓解的疾病,因为没有"择时"可言,所以还是应该立刻救治。

## 42、推波助澜，因势利导

因势利导是指顺着事物发展的趋势，而加以引导和推动，使之出现最好的结果。这个科学方法在很多学术领域都有运用，在军事和医学中更为突出。《史记·孙子 吴起列传》说："善战者，因其势而利导之。"即是说聪明的军事家临战时，必定要详细研究敌我双方之态势，并顺着这种态势发展的趋势，加以引导和推动，使之向有利于我的方面发展，从而以最小的代价换取最大的胜利。

医生的治疗工作，同样要根据病人体内的正气和致病的邪气敌我双方之态势，而采取适宜的治疗措施加以引导和推动，才能最小限度地损伤正气，最快捷、最完全地驱除病邪。清代医学家吴鞠通在其所著《温病条辨》中说："逐邪者，随其性而宣泄之，就其近而引导之。"即要想驱除病邪，就应该根据病邪的性质、特点及其所在部位，给以相应的治疗措施。

例如风邪的性质"轻"，有"散"的特点，容易侵犯人体的上部和皮表部，治疗时就应根据其在上、在表、走散之势，而使用疏风、散风、发汗解表的药物，使之从皮毛散出体外。风热感冒，出现头痛、恶风、咽痛、目赤、咳嗽等症状，就可以用银翘散等方药疏散风热，达到治愈目的。人们熟知的荨麻疹，起病迅速，皮肤起痒疹，俗称泛疙瘩，也是由风邪引起，便可以用消风散等方药，将风邪扬散出去。又如湿邪的性质"重"，有下沉的特点，最容易侵犯人体的下部，因此治湿邪引起的疾病，就常使用利尿、渗湿、化湿的方法,使之逐渐消减而达到治疗目的。

此外,邪气进入人体后,还常常发生转移,中医理论中称为"传化",有从表传入里、从里传出表、从上传到下、从下传到上等不同的传变形式。医生临诊时同样应该掌握邪气转移的趋势,而采取相应的治疗方法。

例如食物停留胃中不化,称为宿食,宿食作为病邪常有向下、向上两种趋向。如有下降之势,出现腹部胀满、大便不通、矢气恶臭等症状时,医生就应该顺其下降之势,给予承气汤类泻下的药物,使之从大便排出;有时病邪虽有下降之势,但因病人大肠津液不足而不能顺利下降,此时则不仅要用泻下药物,还应使用麦冬、玄参之类增津液的药物,所谓"增水行舟",起到推波助澜的作用,使宿食顺利地从大便排出而痊愈。如果宿食有向上的趋势,病人恶心欲吐,自觉只有吐出才能爽快,嗳气频作。此时应据其向上之势,使用摧吐的方法,如瓜蒂散等方药,将病邪经过涌吐而排出体外,从而治愈疾病。吴鞠通就针对疾病传化趋势而治疗的方法总结说:"邪气传化,传表传里,因势导之。"

从上述举例可以看出,运用"因势利导"治疗疾病的优点,在于使病邪从最简捷的途径、以最快的速度排出体外;同时又尽可能少地损伤人体正气,从而有利于疾病痊愈与康复。这与军事上运用"因势利导",以求尽量减少我方损失,尽速取得战争胜利的基本思想是完全一致的。

## 43、权衡得失，和解为宜

军事家指挥作战，根据敌我双方的具体情况，有攻坚、困守、围歼、伏击、奔袭、劫营等多种战法。除以武力直接对抗外，战争中也会出现"讲和"一法，而讲和的基本目的，不外最大程度地保存自己。医学上根据千变万化的病症，也确立有多种治法，概括起来共有八种，即汗、吐、下、和、温、清、消、补。其中的"和"，即是和解之法。

病邪在体表的病症，如感冒、疮疡初起、急性水肿等，适宜用服发汗药或药汤渍浴的汗法；凡实邪在上部，如痰涎在胸中、误服毒物在胃脘等，适宜用催吐药或鹅羽探喉等吐法；燥屎、瘀血等邪气在下部，适宜用攻下或破瘀等下法，使之从大便排出；凡阳气不足，以致病人四肢清冷、腹泻清稀等寒证，适宜用温热药的温法治疗；而病人烦热、口渴、面红、尿黄等热证，应该用寒凉性质的药物来清其热邪，即是清法；若人体禀赋不足，或久病之后正气已虚，而出现削瘦、乏力、盗汗、遗精、月经闭止、气短、心慌、头晕、耳聋等，可据其病情，辨明究竟是阴虚、阳虚、气虚、血虚，还是五脏之虚，而分别采用不同的方药以补其虚，皆属补法之类；病人有食积不化，以致食欲不振、恶心、腹胀、口臭、大便不畅，或者气血凝滞形成肿块病，前者可用消食导滞的药"消"之，后者则使用行气活血、软坚散结的药物，使之"消"散，皆属消法之类。

以上七法，可以广泛适用于表、里、寒、热、虚、实多种病证。但还有一类疾病，既不在表，又不在里，而居于人体的"半表半里"之间；既有正气偏虚，又有邪气未去而为虚实两兼之证。此类病用上述七法都不能起到真正治疗作用，或者药物不能抵达病所，或者补正气即碍邪气，或者泻邪气即伤正气，真可谓"进退两难"。鉴于此种情势，医学家们借助军事上"讲和"之法，创造了"和解"法。使邪气逐渐化解，使正气逐渐平复，使气血调和而疾病痊愈。

例如人患感冒数天之后，咳嗽、头痛、恶寒、发热等症状基本除去，但却出现口苦、咽干、头晕、目眩、阵冷阵热而体温基本正常、恶心欲吐、不思饮食、心烦胸闷等一系列症状。此时中医有一个明确的诊断，叫做"少阳病"，认为这是由于邪气在半表半里所致。也有一个有效药方，叫小柴胡汤，而这小柴胡汤就是典型的"和解"方剂。

当然，中医的和解法不是真正地与邪气讲和，其目的是去除全部病邪，妥善保护好正气。

图书在版编目(CIP)数据

孙子兵法与养生治病:英、汉/吴如嵩等著．－北京：
新世界出版社,1997
ISBN 7-80005-376-8

Ⅰ.孙… Ⅱ.吴… Ⅲ.①孙子兵法-应用-养生(中医)-
英、汉②孙子兵法-应用-保健-英、汉 Ⅳ.R161

中国版本图书馆 CIP 数据核字（97）第 17787 号

| 翻　　译 | 李　斌 |
|---|---|
|  | 佘端志 |
| 译　　审 | 王燕娟 |
| 图书主编 | 吴显林 |
| 编　　审 | 陈士樾 |
| 编　　辑 | 石　岭 |
| 封面设计 | 李士伋 |
| 摄　　影 | 陈宗烈 |
| 责任校对 | 章　弟 |

《孙子兵法与养生治病》

吴如嵩　黄　英　著
王洪图

※

新世界出版社出版
（北京百万庄路 24 号）
邮政编码 100037
外文印刷厂印刷
中国国际图书贸易总公司发行
北京邮政信箱第 399 号　北京车公庄西路 35 号
邮政编码 100044
1997 年第 1 版(英、汉)　1997 年第 1 次印刷
ISBN 7-80005-376-8
05000
17-EC-3016P